SpringerBriefs in Electrical and Computer Engineering

Control, Automation and Robotics

Series editors

Tamer Başar

Antonio Bicchi

Miroslav Krstic

More information about this series at http://www.springer.com/series/10198

Murat Arcak · Chris Meissen
Andrew Packard

Networks of Dissipative Systems

Compositional Certification of Stability, Performance, and Safety

 Springer

Murat Arcak
Department of Electrical Engineering
  and Computer Sciences
University of California, Berkeley
Berkeley, CA
USA

Andrew Packard
Department of Mechanical Engineering
University of California, Berkeley
Berkeley, CA
USA

Chris Meissen
Department of Mechanical Engineering
University of California, Berkeley
Berkeley, CA
USA

Additional material to this book can be downloaded from http://extras.springer.com.

ISSN 2191-8112   ISSN 2191-8120   (electronic)
SpringerBriefs in Electrical and Computer Engineering
ISSN 2192-6786   ISSN 2192-6794   (electronic)
SpringerBriefs in Control, Automation and Robotics
ISBN 978-3-319-29927-3   ISBN 978-3-319-29928-0   (eBook)
DOI 10.1007/978-3-319-29928-0

Library of Congress Control Number: 2016932862

Printed on acid-free paper

This Springer imprint is published by Springer Nature
The registered company is Springer International Publishing AG Switzerland

Preface

Existing computational tools for control synthesis and verification do not scale well to today's large-scale networked systems. Recent advances, such as sum-of-squares relaxations for polynomial nonnegativity, have made it possible to numerically search for Lyapunov functions and to certify measures of performance; however, these procedures are applicable only to problems of modest size.

In this book we address networks where the subsystems are amenable to standard analytical and computational methods but the interconnection, taken as a whole, is beyond the reach of these methods. To break up the task of certifying network properties into subproblems of manageable size, we make use of *dissipativity* properties which serve as abstractions of the detailed dynamical models of the subsystems. We combine these abstractions to derive network level stability, performance, and safety guarantees in a compositional fashion.

Dissipativity theory, which is fundamental to our approach, is reviewed in Chap. 1 and enriched with sum-of-squares and semidefinite programming techniques, detailed in Appendices A and B respectively.

Chapter 2 derives a stability test for interconnected systems from the dissipativity characteristics of the subsystems. This approach is particularly powerful when one exploits the structure of the interconnection and identifies subsystem dissipativity properties favored by the type of interconnection. We exhibit several such interconnections that are of practical importance, as subsequently demonstrated in Chap. 4 with case studies from biological networks, multiagent systems, and Internet congestion control.

Before proceeding to the case studies, however, in Chap. 3 we point out an obstacle to analyzing subsystems independently of each other: the dissipativity properties must be referenced to the network equilibrium point which depends on all other subsystems. To remove this obstacle we introduce the stronger notion of *equilibrium independent dissipativity*, which requires dissipativity with respect to any point that has the potential to become an equilibrium in an interconnection.

In Chap. 5 we extend the compositional stability analysis tools to performance and safety certification. Performance is defined as a desired dissipativity property for the interconnection, such as a prescribed gain from a disturbance input to a performance output. The goal in safety certification is to guarantee that trajectories do not intersect a set that is deemed unsafe.

Unlike the earlier chapters that use a fixed dissipativity property for each subsystem, in Chap. 6 we combine the stability and performance tests with a simultaneous search over compatible subsystem dissipativity properties. We employ the Alternating Direction Method of Multipliers (ADMM) algorithm, a powerful distributed optimization technique, to decompose and solve this problem. In Chap. 7 we exploit the symmetries in the interconnection structure to reduce the number of decision variables, thereby achieving significant computational savings for interconnections that are rich with symmetries.

In Chap. 8 we define a generalized notion of dissipativity that incorporates more information about a dynamical system than the standard form in Chap. 1. This is achieved by augmenting the system model with a linear system that serves as a virtual filter for the inputs and outputs. This dynamic extension is subsequently related to the frequency domain notion of *integral quadratic constraints* in Chap. 9. We conclude by pointing to further results that are complementary to those presented in the book.

Berkeley, CA, USA   Murat Arcak
January 2016   Chris Meissen
                                                                                          Andrew Packard

Acknowledgments

We thank Ana Ferreira, Erin Summers, George Hines, Laurent Lessard, and Sam Coogan for their contributions to the research summarized here.

The work of the authors was funded in part by the National Science Foundation grant ECCS 1405413, entitled "A Compositional Approach for Performance Certification of Large-Scale Engineering Systems" (program director Dr. Kishan Baheti), and by NASA Grant No. NRA NNX12AM55A, entitled "Analytical Validation Tools for Safety Critical Systems Under Loss-of-Control Conditions" (technical monitor Dr. Christine Belcastro). Any opinions, findings, and conclusions or recommendations expressed in this material are those of the author and do not necessarily reflect the views of the NSF or NASA. We also acknowledge generous support from the FANUC Corporation through the FANUC Chair in Mechanical Engineering.

Contents

Chapter 1
Brief Review of Dissipativity Theory

1.1 Dissipative Systems

Consider the dynamical system

$$\frac{\mathrm{d}}{\mathrm{d}t}x(t) = f(x(t), u(t)) \qquad f(0,0) = 0 \tag{1.1}$$

$$y(t) = h(x(t), u(t)) \qquad h(0,0) = 0 \tag{1.2}$$

with $x(t) \in \mathbb{R}^n$, $u(t) \in \mathbb{R}^m$, $y(t) \in \mathbb{R}^p$, and continuously differentiable mappings $f : \mathbb{R}^n \times \mathbb{R}^m \mapsto \mathbb{R}^n$ and $h : \mathbb{R}^n \times \mathbb{R}^m \mapsto \mathbb{R}^p$. Given the input signal $u(\cdot)$ and initial condition $x(0)$, the solution $x(t)$ of (1.1) generates the output $y(t)$ according to (1.2).

The notion of *dissipativity* introduced by Willems [1] characterizes dynamical systems broadly by how their inputs and outputs correlate. The correlation is described by a scalar-valued *supply rate* $s(u, y)$ the choice of which distinguishes the type of dissipativity (Fig. 1.1).

Definition 1.1 The system (1.1)–(1.2) is **dissipative** with respect to a **supply rate** $s(u, y)$ if there exists $V : \mathbb{R}^n \mapsto \mathbb{R}$ such that $V(0) = 0$, $V(x) \geq 0 \; \forall x$, and

$$V(x(\tau)) - V(x(0)) \leq \int_0^\tau s(u(t), y(t))\mathrm{d}t \tag{1.3}$$

for every input signal $u(\cdot)$ and every $\tau \geq 0$ in the interval of existence of the solution $x(t)$. $V(\cdot)$ is called a **storage function**.

© The Author(s) 2016
M. Arcak et al., *Networks of Dissipative Systems*,
SpringerBriefs in Control, Automation and Robotics,
DOI 10.1007/978-3-319-29928-0_1

Fig. 1.1 Dissipativity
characterizes a dynamical
system with a *supply rate*
$s(u, y)$ that describes how
the inputs and outputs
correlate, and an
accompanying *storage
function* $V(\cdot)$

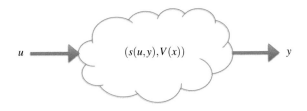

This definition implies that the integral of the supply rate $s(u(t), y(t))$ along the
trajectories is nonnegative when $x(0) = 0$ and lower bounded by the offset $-V(x(0))$
otherwise. Thus, the system favors a positive sign for $s(u(t), y(t))$ when averaged
over time.

Important types of dissipativity are discussed below.

Finite L_2 gain: $s(u, y) = \gamma^2 |u|^2 - |y|^2 \quad \gamma > 0$

We denote by L_2^m the space of functions $u : [0, \infty) \to \mathbb{R}^m$ with finite energy

$$\|u\|_2^2 = \int_0^\infty |u(t)|^2 dt \tag{1.4}$$

where $|\cdot|$ is the Euclidean norm in $\mathbb{R}^m$ and $\|\cdot\|_2$ is the L_2 norm. Note from (1.3)
that

$$-V(x(0)) \le V(x(\tau)) - V(x(0)) \le \gamma^2 \int_0^\tau |u(t)|^2 dt - \int_0^\tau |y(t)|^2 dt$$

$$\Rightarrow \int_0^\tau |y(t)|^2 dt \le \gamma^2 \int_0^\tau |u(t)|^2 dt + V(x(0)).$$

Taking square roots of both sides and applying the inequality $\sqrt{a^2 + b^2} \le |a| + |b|$
to the right-hand side, we get

$$\sqrt{\int_0^\tau |y(t)|^2 dt} \le \gamma \sqrt{\int_0^\tau |u(t)|^2 dt} + \sqrt{V(x(0))}.$$

This means that the L_2 norm $\|y\|_2$ is bounded by $\gamma \|u\|_2$, plus an offset term due to
the initial condition. Thus γ serves as an L_2 gain for the system.

Passivity: $s(u, y) = u^T y$

With this choice of supply rate, (1.3) implies

$$\int_0^\tau u(t)^T y(t) dt \ge -V(x(0)) \tag{1.5}$$

which favors a positive sign for the inner product of $u(t)$ and $y(t)$. Periods of time
when $u(t)^T y(t) < 0$ must be outweighed by those when $u(t)^T y(t) > 0$.

Output strict passivity: $s(u, y) = u^T y - \varepsilon |y|^2$ $\varepsilon > 0$

This supply rate tightens the passivity condition (1.5) as:

$$\int_0^\tau u(t)^T y(t) dt \geq -V(x(0)) + \underbrace{\varepsilon \int_0^\tau |y(t)|^2 dt}_{\geq 0} .$$

In addition, output strict passivity implies an L_2 gain of $\gamma = 1/\varepsilon$ because a completion of squares argument gives

$$u^T y - \frac{1}{\gamma} y^T y \leq \frac{\gamma}{2} u^T u - \frac{1}{2\gamma} y^T y = \frac{1}{2\gamma} (\gamma^2 |u|^2 - |y|^2). \qquad (1.6)$$

Then the storage function $2\gamma V(\cdot)$ yields the L_2 gain supply rate $\gamma^2 |u|^2 - |y|^2$.

1.2 Graphical Interpretation

For a memoryless system

$$y(t) = h(u(t))$$

we take the storage function in (1.3) to be zero and interpret dissipativity as the static inequality

$$s(u, h(u)) \geq 0 \qquad \forall u \in \mathbb{R}^m \qquad (1.7)$$

which characterizes the maps $h(\cdot)$ that are dissipative with supply rate $s(\cdot, \cdot)$.

For example, a scalar function $h(\cdot)$ is passive if $uh(u) \geq 0$ for all u, which means that the graph of $h(\cdot)$ lies in the first and third quadrants as in Fig. 1.2 (left).

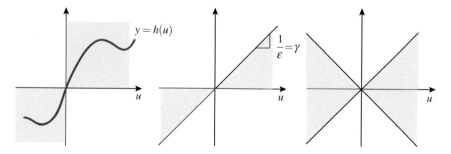

Fig. 1.2 The graph of a passive static nonlinearity $h(\cdot)$ lies in the first and third quadrants (*left*). Output strict passivity confines $h(\cdot)$ to the narrower sector (*middle*) and a gain bound γ corresponds to the sector upper and lower bounded by the lines $\pm \gamma u$ (*right*)

Likewise, the sector in the middle represents the output strict passivity supply rate $s(u, y) = uy - \varepsilon y^2$, $\varepsilon > 0$, and the sector on the right represents the finite gain supply rate $s(u, y) = \gamma^2 u^2 - y^2$.

1.3   Differential Characterization of Dissipativity

When the storage function $V(\cdot)$ is continuously differentiable, the dissipation inequality (1.3) is equivalent to

$$\nabla V(x)^T f(x, u) \leq s(u, h(x, u)) \qquad \forall x \in \mathbb{R}^n, \forall u \in \mathbb{R}^m. \tag{1.8}$$

Thus, to verify dissipativity we search for a $V(\cdot)$ satisfying $V(0) = 0$, $V(x) \geq 0$, and (1.8) for all x and u.

As an illustration, suppose we wish to prove passivity of the system

$$\frac{\mathrm{d}}{\mathrm{d}t} x(t) = f_0(x(t)) + g(x(t))u(t)$$
$$y(t) = h(x(t))$$

which is a special case of (1.1)–(1.2) with $f(x, u) = f_0(x) + g(x)u$ affine in u, and $h(x, u) = h(x)$ independent of u. Then (1.8) becomes

$$\nabla V(x)^T f_0(x) + \nabla V(x)^T g(x)u \leq h(x)^T u \quad \forall x \in \mathbb{R}^n, \forall u \in \mathbb{R}^m \tag{1.9}$$

which is equivalent to

$$\nabla V(x)^T f_0(x) \leq 0 \qquad \nabla V(x)^T g(x) = h^T(x) \qquad \forall x \in \mathbb{R}^n. \tag{1.10}$$

The inequality in (1.10) follows from (1.9) when $u = 0$. To see how the equality follows suppose, to the contrary, there exists an x for which $\nabla V(x)^T g(x) - h^T(x) \neq 0$. Then we can select a u such that $(\nabla V(x)^T g(x) - h^T(x))u$ is positive and large enough to contradict (1.9).

Similar arguments show that output strict passivity is equivalent to

$$\nabla V(x)^T f_0(x) \leq -\varepsilon h(x)^T h(x) \quad \nabla V(x)^T g(x) = h^T(x) \quad \forall x \in \mathbb{R}^n. \tag{1.11}$$

Example 1.1 Consider the scalar system

$$\frac{\mathrm{d}x(t)}{\mathrm{d}t} = f_0(x(t)) + u(t), \quad y(t) = h(x(t)), \quad u(t), x(t), y(t) \in \mathbb{R} \tag{1.12}$$

where $h(\cdot)$ satisfies $xh(x) \geq 0$ for all x, as in Fig. 1.2 (left). For this system the equality in (1.11) is

$$\frac{dV(x)}{dx} = h(x)$$

whose solution subject to $V(0) = 0$ is

$$V(x) = \int_0^x h(z)dz. \tag{1.13}$$

Furthermore $V(x) \geq 0$ because $h(z)$ and dz have equal signs (positive when the limit of integration is $x > 0$ and negative when $x < 0$).

The inequality condition in (1.11) is then

$$h(x)(f_0(x) + \varepsilon h(x)) \leq 0$$

which is equivalent to

$$x(f_0(x) + \varepsilon h(x)) \leq 0 \tag{1.14}$$

since $xh(x) \geq 0$. Thus, we conclude passivity when (1.14) holds with $\varepsilon = 0$ and output strict passivity when (1.14) holds with $\varepsilon > 0$.

For an *integrator*, where $f_0(x) \equiv 0$ and $h(x) = x$, (1.14) becomes $\varepsilon x^2 \leq 0$ which holds only with $\varepsilon = 0$. Thus we have passivity but not output feedback passivity.

Example 1.2 Consider the second-order model

$$\frac{dx_1(t)}{dt} = x_2(t)$$
$$\frac{dx_2(t)}{dt} = -kx_2(t) - \phi'(x_1(t)) + u(t)$$
$$y(t) = x_2(t)$$

where $\phi'(\cdot)$ is the derivative of a continuously differentiable and nonnegative function $\phi(\cdot)$ satisfying $\phi(0) = 0$. We interpret x_1 as position, x_2 as velocity, u as force, $k \geq 0$ as damping coefficient, and $\phi(x_1)$ as potential energy of a mechanical system.

For this system the equality condition $\nabla V(x)^T g(x) = h^T(x)$ becomes:

$$\frac{\partial V(x_1, x_2)}{\partial x_2} = x_2.$$

Thus we restrict the storage function to be of the form:

$$V(x_1, x_2) = V_1(x_1) + \frac{1}{2}x_2^2$$

and examine the inequality condition $\nabla V(x)^T f_0(x) \leq 0$. We have

$$
\nabla V(x)^T f_0(x) = \frac{dV_1(x_1)}{dx_1} x_2 + x_2 \left(-kx_2 - \phi'(x_1)\right)
$$
$$
= -kx_2^2 + x_2 \left(\frac{dV_1(x_1)}{dx_1} - \phi'(x_1)\right).
$$

The choice $V_1(x_1) = \phi(x_1)$ ensures $\nabla V(x)^T f_0(x) = -kx_2^2 = -kh(x)^2$ which proves passivity when $k = 0$ and output strict passivity when $k > 0$.

   The resulting storage function $V(x_1, x_2) = \phi(x_1) + \frac{1}{2}x_2^2$ is the sum of potential and kinetic energy terms, and $u(t)y(t)$ =force×velocity may be interpreted as the power supplied to the system. The definition of dissipativity (1.3) is thus consistent with the physical notion of energy storage, and dissipation when damping is present.

1.4 Linear Systems

A linear system is dissipative with respect to a quadratic supply rate *if and only if* (1.8) is satisfied with a quadratic storage function [2]. Thus, given the system

$$
\frac{d}{dt}x(t) = Ax(t) + Bu(t) \tag{1.15}
$$
$$
y(t) = Cx(t) + Du(t), \tag{1.16}
$$

$A \in \mathbb{R}^{n \times n}, B \in \mathbb{R}^{n \times m}, C \in \mathbb{R}^{p \times n}, D \in \mathbb{R}^{p \times m}$, and the quadratic supply rate

$$
s(u, y) = \begin{bmatrix} u \\ y \end{bmatrix}^T X \begin{bmatrix} u \\ y \end{bmatrix} = \begin{bmatrix} u \\ Cx + Du \end{bmatrix}^T X \begin{bmatrix} u \\ Cx + Du \end{bmatrix} \tag{1.17}
$$

where $X = X^T \in \mathbb{R}^{(m+p) \times (m+p)}$, we restrict our search to a storage function of the form $V(x) = \frac{1}{2}x^T Px$ where $P \in \mathbb{R}^{n \times n}$ is positive semidefinite. Then (1.8) becomes

$$
\frac{1}{2}(Ax + Bu)^T Px + \frac{1}{2}x^T P(Ax + Bu) \leq \begin{bmatrix} u \\ Cx + Du \end{bmatrix}^T X \begin{bmatrix} u \\ Cx + Du \end{bmatrix} \tag{1.18}
$$

$\forall x \in \mathbb{R}^n, \forall u \in \mathbb{R}^m$, which is equivalent to the matrix inequality

$$
\frac{1}{2} \begin{bmatrix} A^T P + PA & PB \\ B^T P & 0 \end{bmatrix} \leq \begin{bmatrix} 0 & I \\ C & D \end{bmatrix}^T X \begin{bmatrix} 0 & I \\ C & D \end{bmatrix}. \tag{1.19}
$$

As a special case, for the passivity supply rate $s(u, y) = u^T y$, where

$$X = \begin{bmatrix} 0 & \frac{1}{2}I \\ \frac{1}{2}I & 0 \end{bmatrix},$$

(1.19) with $D = 0$ becomes

$$\begin{bmatrix} A^T P + PA & PB - C^T \\ B^T P - C & 0 \end{bmatrix} \leq 0. \tag{1.20}$$

This inequality can hold only if the off-diagonal block is zero, $PB - C^T = 0$, hence

$$A^T P + PA \leq 0 \quad PB = C^T \tag{1.21}$$

is equivalent to (1.20) and parallels the condition (1.10) above for the nonlinear case.

Example 1.3 We show that the second order system with

$$A = \begin{bmatrix} 0 & 1 \\ -\ell & -k \end{bmatrix} \quad B = \begin{bmatrix} 0 \\ \gamma \end{bmatrix} \quad C = \begin{bmatrix} \mu & 1 \end{bmatrix} \quad D = 0, \tag{1.22}$$

where $\ell > 0$ and $\gamma > 0$, is passive if and only if $k \geq \mu \geq 0$.

To see the necessity note that the constraint $PB = C^T$ restricts P to the form

$$P = \frac{1}{\gamma} \begin{bmatrix} q & \mu \\ \mu & 1 \end{bmatrix} \tag{1.23}$$

and the constraint $A^T P + PA \leq 0$ restricts the diagonal entries of

$$A^T P + PA = -\frac{1}{\gamma} \begin{bmatrix} 2\mu\ell & \mu k + \ell - q \\ \mu k + \ell - q & 2(k - \mu) \end{bmatrix} \tag{1.24}$$

by $\mu\ell \geq 0$ and $k - \mu \geq 0$; hence $k \geq \mu \geq 0$.

To see the sufficiency, suppose $k \geq \mu \geq 0$ and select $q = \mu k + \ell$ in (1.23). Then $A^T P + PA \leq 0$ follows trivially from (1.24) and $P > 0$ follows because $q = \mu k + \ell \geq \mu^2 + \ell > \mu^2$ guarantees the determinant of (1.23) is positive.

The arguments above also imply that there exists $P = P^T > 0$ satisfying

$$A^T P + PA < 0 \quad PB = C^T, \tag{1.25}$$

that is (1.21) with strict inequality, if and only if $k > \mu > 0$. In particular, the strict inequality in (1.25) allows us to find $\varepsilon > 0$ such that $A^T P + PA + 2\varepsilon C^T C \leq 0$ which implies (1.19) with

$$X = \begin{bmatrix} 0 & \frac{1}{2} \\ \frac{1}{2} & -\varepsilon \end{bmatrix}. \tag{1.26}$$

Thus $k > \mu > 0$ guarantees output strict passivity.

Example 1.4 Consider a linear single-input single-output system of the form

$$\hat{A} = \begin{bmatrix} A & 0 \\ 0 & A_0 \end{bmatrix}, \quad \hat{B} = \begin{bmatrix} B \\ 0 \end{bmatrix}, \quad B \neq 0, \quad \hat{C} = \begin{bmatrix} C & C_0 \end{bmatrix}, \quad \hat{D} = 0$$

where the subsystem governed by A_0 represents uncontrollable dynamics. If the rest of the system admits a matrix $P = P^T > 0$ satisfying (1.25) and all eigenvalues of A_0 have negative real parts, then there exists $\hat{P} = \hat{P}^T > 0$ satisfying

$$\hat{A}^T \hat{P} + \hat{P} \hat{A} < 0 \quad \hat{P}\hat{B} = \hat{C}^T. \tag{1.27}$$

We leave it to the reader to prove this claim with a matrix of the form

$$\hat{P} = \begin{bmatrix} P & R \\ R^T & \gamma P_0 \end{bmatrix}$$

where $P_0 = P_0^T > 0$ satisfies $A_0^T P_0 + P_0 A_0 < 0$, R must be selected appropriately, and $\gamma > 0$ must be selected large enough to ensure $\hat{P} > 0$ and $\hat{A}^T \hat{P} + \hat{P} \hat{A} < 0$.

1.5 Numerical Certification of Dissipativity

Note that (1.19) is a standard linear matrix inequality (LMI) feasibility problem in $P \geq 0$ and X, and can be solved with convex optimization packages such as CVX [3] or YALMIP [4]. These packages formulate the problem as a semidefinite program (SDP) and then call appropriate solvers. Appendix B reviews recent advances that improve the computational efficiency of SDP solvers, including in the case where no *strictly* feasible solutions exists. An example of this case is passivity certification where (1.20) above can be at most semidefinite.

When $f(x, u)$ and $h(x, u)$ in (1.1)–(1.2) are polynomials, dissipativity can be certified using sum-of-squares (SOS) programming. Let $\mathbb{R}[x]$ be the set of polynomials in x and $\Sigma[x] \subset \mathbb{R}[x]$ be the subset of all SOS polynomials. A polynomial system is dissipative with respect to a polynomial supply rate, $s(u, h(x, u)) \in \mathbb{R}[x, u]$, if there exists a function $V(\cdot)$ satisfying the SOS feasibility problem

$$V(x) \in \Sigma[x] \tag{1.28}$$

$$-\nabla V(x)^T f(x, u) + s(u, h(x, u)) \in \Sigma[x, u]. \tag{1.29}$$

The constraint $V(0) = 0$ is enforced by excluding constant terms in the choice of the monomials that constitute $V(x)$.

As shown in Appendix A, SOS feasibility problems such as (1.28)–(1.29) can be relaxed to SDPs and solved with standard software packages.

Unlike linear systems where there is no loss in restricting the search to quadratic storage functions, (1.28)–(1.29) is only a sufficient condition for dissipativity since SOS polynomials form a strict subset of all nonnegative polynomials. Furthermore, the degree of the storage function $V(\cdot)$ must be limited to prevent the problem from becoming computationally intractable.

1.6   Using Dissipativity for Reachability and Stability

A common approach to studying input/output properties is to treat dynamical systems as operators mapping inputs to outputs in appropriate function spaces, as presented in [5]. Unlike this approach, dissipativity theory allows us to derive input/output properties from a state space model and to establish bounds on the state trajectories using bounds on the storage function. We illustrate the latter by deriving reachability bounds and Lyapunov stability properties with appropriate choices of supply rates.

L_2 *reachability*: $s(u, y) = |u|^2$

This supply rate implies

$$V(x(\tau)) \le \int_0^\tau |u(t)|^2 dt + V(x(0)).$$

Hence, if $\|u\|_2^2 \le \beta$, then $V(x(\tau)) \le \beta + V(x(0))$ for all $\tau \ge 0$, which means that trajectories starting in the sublevel set

$$\mathcal{V}_\alpha = \{x : V(x) \le \alpha\}$$

remain in the sublevel set $\mathcal{V}_{\alpha+\beta}$, as depicted in Fig. 1.3 (left).

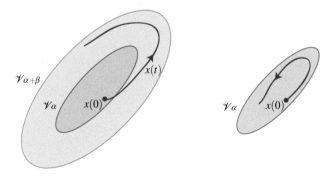

Fig. 1.3   Dissipativity with the L_2 reachability supply rate $s(u, y) = |u|^2$ and storage function $V(\cdot)$ ensures that trajectories starting in the sublevel set $\mathcal{V}_\alpha = \{x : V(x) \le \alpha\}$ remain in the enlarged sublevel set $\mathcal{V}_{\alpha+\beta}$ for all inputs u such that $\|u\|_2^2 \le \beta$ (*left*). In particular, when $u(t) \equiv 0$, trajectories starting in $\mathcal{V}_\alpha$ remain in $\mathcal{V}_\alpha$ thereafter (*right*)

Lyapunov stability

When $u(t) \equiv 0$, a dissipative system whose supply rate $s(u, y)$ is such that

$$s(0, 0) = 0, \quad s(0, y) \leq 0 \quad \forall y \in \mathbb{R}^p, \tag{1.30}$$

guarantees that trajectories starting in the sublevel set $\mathcal{V}_\alpha$ remain in $\mathcal{V}_\alpha$, because

$$V(x(\tau)) \leq \int_0^\tau \underbrace{s(0, y(t))}_{\leq 0}\, dt + V(x(0)) \leq V(x(0)).$$

The L_2 reachability supply rate above as well as those discussed in Sect. 1.1 satisfy (1.30).

If, in addition, $V(\cdot)$ is positive definite ($V(0) = 0$, $V(x) > 0$ for $x \neq 0$) then the storage function serves as a Lyapunov function and certifies stability for the equilibrium $x = 0$ of the system (1.1) with $u(t) \equiv 0$:

$$\frac{d}{dt} x(t) = f(x(t), 0) \qquad f(0, 0) = 0.$$

The positive definiteness of $V(\cdot)$ ensures that the sublevel sets $\mathcal{V}_\alpha$ are compact for sufficiently small α; therefore, trajectories starting close to $x = 0$ remain close as in Fig. 1.3 (right)—the core principle in Lyapunov stability theory [6].

If $V(\cdot)$ is *radially unbounded*, that is, $V(x) \to \infty$ as $|x| \to \infty$ along any path in $\mathbb{R}^n$, then $\mathcal{V}_\alpha$ is compact no matter how large α; therefore all trajectories are bounded and the stability property is *global*.

Asymptotic stability can be established by further examining the right-hand side of (1.8) with $u = 0$:

$$\nabla V(x)^T f(x, 0) \leq s(0, h(x, 0)) \qquad \forall x \in \mathbb{R}^n. \tag{1.31}$$

If (1.30) holds with strict inequality for $y \neq 0$, then the right-hand side of (1.31) vanishes when $h(x, 0) = 0$ and is strictly negative otherwise. Thus, we can appeal to the *Invariance Principle* [6] which states that, if the only solution satisfying $h(x(t), 0) = 0$ for all t is $x(t) = 0$, then $x = 0$ is asymptotically stable.

The following chapters compose Lyapunov functions for interconnections using the dissipativity properties of the subsystems. We deemphasize the type of stability achieved (local or global, asymptotic or not) as this can be determined with standard techniques such as the ones alluded to above. Instead, we focus on how a Lyapunov function can be composed in the first place—a task hindered in large networks by the state dimension and the need for explicit knowledge of the equilibrium.

Since this first chapter is foundational for the rest of the book, we include true/false questions in Appendix D for readers who are new to the subject. For further details on dissipativity theory we refer the readers to the monographs [7, 8].

References

1. Willems, J.: Dissipative dynamical systems, Part I: general theory. Arch. Ration. Mech. Anal. **45**, 321–351 (1972)
2. Willems, J.: Dissipative dynamical systems, Part II: linear systems with quadratic supply rates. Arch. Ration. Mech. Anal. **45**, 352–393 (1972)
3. Grant, M., Boyd, S.: CVX: Matlab software for disciplined convex programming. http://cvxr.com/cvx (2014)
4. Löfberg, J.: Yalmip : A toolbox for modeling and optimization in MATLAB. In: Proceedings of the CACSD Conference, Taipei, Taiwan (2004)
5. Desoer, C., Vidyasagar, M.: Feedback systems: input-output properties. In: Society for Industrial and Applied Mathematics, Philadelphia (2009). Originally published by Academic Press, New York (1975)
6. Khalil, H.: Nonlinear Systems, 3rd edn. Prentice Hall, Upper Saddle River (2002)
7. Brogliato, B., Lozano, R., Maschke, B., Egeland, O.: Dissipative Systems Analysis and Control: Theory and Applications. Communications and Control Engineering. Springer, London (2007)
8. van der Schaft, A.J.: $\mathscr{L}_2$-gain and Passivity Techniques in Nonlinear Control, 2nd edn. Springer, New York (2000)

Chapter 2
Stability of Interconnected Systems

Consider the interconnection in Fig. 2.1 where each subsystem G_i, $i = 1, \ldots, N$, is described by

$$\frac{\mathrm{d}}{\mathrm{d}t} x_i(t) = f_i(x_i(t), u_i(t)) \tag{2.1}$$

$$y_i(t) = h_i(x_i(t), u_i(t)) \tag{2.2}$$

with $x_i(t) \in \mathbb{R}^{n_i}$, $u_i(t) \in \mathbb{R}^{m_i}$, $y_i(t) \in \mathbb{R}^{p_i}$, $f_i(0, 0) = 0$, $h_i(0, 0) = 0$.

The static matrix M defines the coupling of these subsystems: the input u_i to G_i depends on the outputs y_j of other subsystems by

$$u = My \tag{2.3}$$

where $u = [u_1^T \cdots u_N^T]^T$ and $y = [y_1^T \cdots y_N^T]^T$. We assume that the interconnection is well-posed; that is, upon the substitution $y_i = h_i(x_i, u_i)$ the Eq. (2.3) admits a unique solution for u as a function x.

2.1 Compositional Stability Certification

Our goal is to derive a bottom-up stability test using dissipativity properties and the interconnection structure of the subsystems. Dissipativity serves as an abstraction of the subsystem models (Fig. 1.1) and allows us to study interconnections whose combined dynamical equations are too large to analyze directly. The use of input/output properties and interconnection matrices for network stability tests dates back to the early Refs. [1, 2].

© The Author(s) 2016
M. Arcak et al., *Networks of Dissipative Systems*,
SpringerBriefs in Control, Automation and Robotics,
DOI 10.1007/978-3-319-29928-0_2

Fig. 2.1 An interconnection
of subsystems $G_1, \ldots, G_N$.
The inputs depend on the
outputs of other subsystems
by $u = My$ where M is a
static matrix

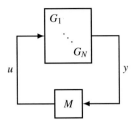

We assume each subsystem is dissipative with a positive definite, continuously
differentiable storage function $V_i(\cdot)$ and a quadratic supply rate:

$$s_i(u_i, y_i) = \begin{bmatrix} u_i \\ y_i \end{bmatrix}^T X_i \begin{bmatrix} u_i \\ y_i \end{bmatrix} = \begin{bmatrix} u_i \\ y_i \end{bmatrix}^T \begin{bmatrix} X_i^{11} & X_i^{12} \\ X_i^{21} & X_i^{22} \end{bmatrix} \begin{bmatrix} u_i \\ y_i \end{bmatrix} \qquad (2.4)$$

where X_i^{jk}, $j, k \in \{1, 2\}$, are conformal block partitions of X_i. We then search for a
weighted sum of storage functions

$$V(x) = p_1 V_1(x_1) + \cdots + p_N V_N(x_N) \quad p_i > 0, \ i = 1, \ldots, N \qquad (2.5)$$

that serves as a Lyapunov function for the interconnection. To this end we ask that
the right-hand side of the inequality

$$\sum_{i=1}^{N} p_i \nabla V_i(x_i)^T f_i(x_i, u_i) \le \sum_{i=1}^{N} p_i \begin{bmatrix} u_i \\ y_i \end{bmatrix}^T X_i \begin{bmatrix} u_i \\ y_i \end{bmatrix} \qquad (2.6)$$

be negative semidefinite in y when u is eliminated with the substitution $u = My$.
Rewriting the right-hand side of (2.6) as

$$\underbrace{\begin{bmatrix} u_1 \\ \vdots \\ u_N \\ y_1 \\ \vdots \\ y_N \end{bmatrix}^T \begin{bmatrix} p_1 X_1^{11} & & & p_1 X_1^{12} & & \\ & \ddots & & & \ddots & \\ & & p_N X_N^{11} & & & p_N X_N^{12} \\ p_1 X_1^{21} & & & p_1 X_1^{22} & & \\ & \ddots & & & \ddots & \\ & & p_N X_N^{21} & & & p_N X_N^{22} \end{bmatrix} \begin{bmatrix} u_1 \\ \vdots \\ u_N \\ y_1 \\ \vdots \\ y_N \end{bmatrix}}_{\triangleq \mathbf{X}(p_1 X_1, \ldots, p_N X_N)} \qquad (2.7)$$

$$= y^T \begin{bmatrix} M \\ I \end{bmatrix}^T \mathbf{X}(p_1 X_1, \ldots, p_N X_N) \begin{bmatrix} M \\ I \end{bmatrix} y$$

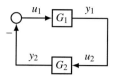

Fig. 2.2 When M is as in (2.9), $u = My$ describes a negative feedback interconnection of two subsystems where $u_1 = -y_2$ and $u_2 = y_1$

we obtain the following stability criterion:

Proposition 2.1 *If there exist $p_i > 0$, $i = 1, \ldots, N$, such that*

$$\begin{bmatrix} M \\ I \end{bmatrix}^T \mathbf{X}(p_1 X_1, \ldots, p_N X_N) \begin{bmatrix} M \\ I \end{bmatrix} \leq 0 \tag{2.8}$$

where $\mathbf{X}(p_1 X_1, \ldots, p_N X_N)$ is as defined in (2.7), then $x = 0$ is stable for the interconnected system (2.1)–(2.3) and (2.5) is a Lyapunov function.

For memoryless subsystems of the form $y_i(t) = h_i(u_i(t))$ we take the corresponding storage function in (2.5) to be zero.

Asymptotic stability requires additional assumptions, such as strict inequality in (2.8) accompanied with an argument that $x(t) = 0$ is the only solution satisfying $h_i(x_i(t), 0) = 0$, $i = 1, \ldots, N$, for all t.

Note that (2.8) is a linear matrix inequality (LMI) and the search for $p_i > 0$ satisfying this inequality can be performed with convex optimization packages [3, 4].

Below we assume each subsystem is single-input single-output and specialize the LMI (2.8) to particular types of dissipativity. This allows us to derive analytical feasibility conditions for special interconnection matrices M. Of particular interest is

$$M = \begin{bmatrix} 0 & -1 \\ 1 & 0 \end{bmatrix} \tag{2.9}$$

which describes the negative feedback loop of two subsystems (Fig. 2.2), commonly studied in control theory.

2.2 Small Gain Criterion

Suppose each subsystem possesses a finite L_2 gain; that is, the supply rate in (2.4) is

$$X_i = \begin{bmatrix} \gamma_i^2 & 0 \\ 0 & -1 \end{bmatrix}.$$

Defining $P \triangleq \mathrm{diag}(p_1, \ldots, p_N)$ and $\Gamma \triangleq \mathrm{diag}(\gamma_1, \ldots, \gamma_N)$ we get

$$\mathbf{X}(p_1 X_1, \ldots, p_N X_N) = \begin{bmatrix} \Gamma P \Gamma & 0 \\ 0 & -P \end{bmatrix}$$

and (2.8) becomes

$$(\Gamma M)^T P (\Gamma M) - P \leq 0. \tag{2.10}$$

Thus a diagonal matrix $P > 0$ satisfying this LMI certifies the stability of the interconnection.

When M is as in (2.9), the LMI (2.10) becomes

$$\begin{bmatrix} p_2 \gamma_2^2 & 0 \\ 0 & p_1 \gamma_1^2 \end{bmatrix} - \begin{bmatrix} p_1 & 0 \\ 0 & p_2 \end{bmatrix} \leq 0$$

which consists of two simultaneous inequalities, $p_2 \gamma_2^2 \leq p_1$ and $p_1 \gamma_1^2 \leq p_2$. We rewrite them as

$$\gamma_2^2 \leq \frac{p_1}{p_2} \leq \frac{1}{\gamma_1^2}$$

and note that such $p_1 > 0$ and $p_2 > 0$ exist if and only if $\gamma_2^2 \leq \frac{1}{\gamma_1^2}$, that is

$$\gamma_1 \gamma_2 \leq 1. \tag{2.11}$$

This condition restricts the loop gain in Fig. 2.2 and is known as a "small gain" criterion.

Note that the derivation above yields the same condition, (2.11), when adapted to the *positive* feedback interconnection where

$$M = \begin{bmatrix} 0 & 1 \\ 1 & 0 \end{bmatrix}.$$

This means that the small gain criterion is oblivious to the feedback sign.

2.3 Passivity Theorem

We now specialize Proposition 2.1 to passivity where

$$X_i = \begin{bmatrix} 0 & 1/2 \\ 1/2 & -\varepsilon_i \end{bmatrix} \quad \varepsilon_i \geq 0.$$

With $P \triangleq \mathrm{diag}(p_1, \ldots, p_N)$ and $E \triangleq \mathrm{diag}(\varepsilon_1, \ldots, \varepsilon_N)$ we get

$$\mathbf{X}(p_1 X_1, \ldots, p_N X_N) = \frac{1}{2} \begin{bmatrix} 0 & P \\ P & -2PE \end{bmatrix}$$

which means that (2.8) is equivalent to

$$P(M - E) + (M - E)^T P \le 0 \qquad (2.12)$$

and a diagonal matrix $P > 0$ satisfying this LMI certifies the stability of the interconnected system.

From matrix Hurwitz stability theory, (2.12) with $P > 0$ implies that all eigenvalues of $M - E$ are within the closed left half-plane. Thus, if $M - E$ has an eigenvalue with a strictly positive real part, there is no $P > 0$ satisfying (2.12). However, we cannot confirm the feasibility of (2.12) with a *diagonal* $P > 0$ from the eigenvalues alone.

Below we exhibit practically important classes of interconnection structures for which (2.12) admits a diagonal solution $P > 0$.

2.3.1 Skew Symmetric Interconnections

The stability criterion (2.12) holds trivially with $P = I$ when M is skew symmetric:

$$M + M^T = 0.$$

There is no restriction on the number or the gains of subsystems, which makes passivity ideally suited to large-scale systems with a skew symmetric coupling structure.

In Chap. 4 we show that this structure arises naturally in distributed control of vehicle platoons and in Internet congestion control. A simpler example of a skew symmetric interconnection is the negative feedback interconnection of two subsystems (Fig. 2.2) where M is as in (2.9). The stability of this interconnection with passive subsystems is a classical result known as the passivity theorem.

2.3.2 Negative Feedback Cyclic Interconnection

To derive another special case of the stability criterion (2.12), we consider a negative feedback loop of N subsystems where the interconnection matrix is

$$M = \begin{bmatrix} 0 & \cdots & 0 & \delta_1 \\ \delta_2 & 0 & \cdots & 0 \\ \vdots & \ddots & \ddots & \vdots \\ 0 & \cdots & \delta_N & 0 \end{bmatrix} \quad \text{with} \quad \prod_{i=1}^{N} \delta_i = -1. \qquad (2.13)$$

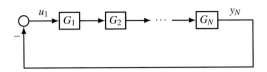

Fig. 2.3 A negative feedback cyclic interconnection of N subsystems. In this example M is as in (2.13) with $\delta_1 = -1, \delta_2 = \cdots = \delta_N = 1$

One such interconnection is shown in Fig. 2.3 where $\delta_1 = -1, \delta_2 = \cdots = \delta_N = 1$.

We prove in Sect. 7.2 that (2.12) admits a diagonal solution $P > 0$ for the class of matrices (2.13) if and only if

$$\prod_{i=1}^{N} \varepsilon_i \geq \cos^N(\pi/N). \tag{2.14}$$

In addition, it was shown in [5] that (2.12) holds with strict inequality if and only if (2.14) is strict.

For $N = 2$ the condition (2.14) recovers the classical passivity theorem: $\cos(\pi/2) = 0$ and passivity ($\varepsilon_i \geq 0$) guarantees stability. For $N \geq 3$, $\cos(\pi/N) > 0$ and (2.14) demands output strict passivity ($\varepsilon_i > 0$).

To compare (2.14) to the small gain criterion, we recall from Sect. 1.1 that output strict passivity implies an L_2 gain of $\gamma_i = 1/\varepsilon_i$ and rewrite (2.14) as

$$\prod_{i=1}^{N} \gamma_i \leq \sec^N(\pi/N) \tag{2.15}$$

where $\sec(\cdot) = 1/\cos(\cdot)$. Unlike the small gain criterion which restricts the feedback loop gain by one, the "secant condition" (2.15) offers the relaxed bound $\sec^N(\pi/N)$ which is equal to 8 when $N = 3$, and decreases asymptotically to one as $N \to \infty$. This sharper bound is due to the output strict passivity assumption which restricts the subsystems further than an L_2 gain property.

Example 2.1 Consider the following model for a *ring oscillator* circuit (Fig. 2.4) that consists of a feedback loop of three inverters:

$$\tau_1 \frac{dx_1(t)}{dt} = -x_1(t) - h_3(x_3(t))$$
$$\tau_2 \frac{dx_2(t)}{dt} = -x_2(t) - h_1(x_1(t)) \tag{2.16}$$
$$\tau_3 \frac{dx_3(t)}{dt} = -x_3(t) - h_2(x_2(t))$$

where $\tau_i = R_i C_i > 0$, $i = 1, 2, 3$, and x_i represent voltages. The functions $h_i(\cdot)$ depend on the inverter characteristics and satisfy

Fig. 2.4 Schematic of a
three-stage ring oscillator
circuit

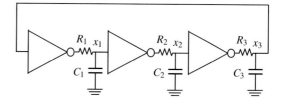

$$h_i(0) = 0, \quad xh_i(x) > 0 \quad \forall x \neq 0, \tag{2.17}$$

as in the commonly used model

$$h_i(x) = \alpha_i \tanh(\beta_i x) \quad \alpha_i > 0, \beta_i > 0. \tag{2.18}$$

We decompose (2.16) into the subsystems

$$G_i: \quad \tau_i \frac{dx_i(t)}{dt} = -x_i(t) + u_i(t) \quad y_i(t) = h_i(x_i(t))$$

interconnected according to $u = My$ where $M \in \mathbb{R}^{3 \times 3}$ is as in (2.13) with $\delta_1 = \delta_2 = \delta_3 = -1$.

Next, we note from (1.14) with $f_0(x) = -x$ that the subsystems are output strictly passive if

$$\varepsilon_i xh_i(x) \leq x^2.$$

This inequality, combined with (2.17), restricts the graph of $h_i(\cdot)$ to the sector in Fig. 1.2 (middle) with slope $\gamma_i = 1/\varepsilon_i$. An example of such a function is (2.18) where $\gamma_i = \alpha_i \beta_i$.

Then, an application of (2.15) with $N = 3$ shows that the equilibrium of the interconnection $x = 0$ is stable when

$$\gamma_1 \gamma_2 \gamma_3 \leq 8 \tag{2.19}$$

and a weighted sum of storage functions, each constructed as in (1.13), serves as a Lyapunov function:

$$V(x) = \sum_{i=1}^{3} p_i \int_0^{x_i} h_i(z)dz.$$

The weights $p_i > 0$ are obtained from the LMI (2.12) which is guaranteed to have a diagonal solution $P > 0$ by (2.19). When the inequality (2.19) is strict we conclude asymptotic stability because (2.12) is negative definite, which means that (2.7) is a negative definite function of y and, further, $y_i = h_i(x_i) = 0 \Rightarrow x_i = 0$ by (2.17).

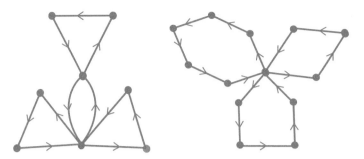

Fig. 2.5 Examples of cactus graphs

When $\tau_1 = \tau_2 = \tau_3$, the secant condition (2.19) is also necessary for stability [5]. Once the loop gain exceeds 8, the equilibrium loses its stability and a limit cycle emerges, hence, the term "ring oscillator."

2.3.3 Extension to Cactus Graphs

To describe a broader interconnection structure that encompasses the cyclic interconnection above, we define an incidence graph for M by directing an edge from vertex j to i if and only if $m_{ij} \neq 0$. This graph is said to be a *cactus graph* if any pair of distinct simple cycles[1] have at most one common vertex, as in the examples of Fig. 2.5.

For matrices M with this structure and $E \triangleq \mathrm{diag}(\varepsilon_1, \ldots, \varepsilon_N) > 0$, a procedure was developed in [6] to determine the range of the entries of M and E for which a diagonal $P > 0$ satisfies (2.12) with strict inequality. This procedure assigns the weight m_{ij}/ε_i to the edge connecting vertex j to i and calculates the gain Γ_c for each cycle $c = 1, \ldots, C$ by multiplying the weights along the cycle. It then restricts the cycle gains according to the specific topology of the graph.

When applied to the subclass of cactus graphs where *all* cycles intersect at one common vertex as in Fig. 2.5 (right), this procedure yields the condition

$$\sum_{c=1}^{C} \alpha_c \Gamma_c < 1 \quad \text{where} \quad \alpha_c = \begin{cases} 1 & \text{if } \Gamma_c > 0 \\ -\cos^{n_c}(\pi/n_c) & \text{if } \Gamma_c < 0 \end{cases} \qquad (2.20)$$

and n_c is the number of edges on cycle c. For a single cycle ($C = 1$) with negative gain $\Gamma < 0$ and N edges, (2.20) becomes

$$\alpha \Gamma = |\Gamma| \cos^N(\pi/N) < 1,$$

thus recovering the strict form of the secant condition.

---

[1] Simple cycles are cycles with no repeated vertices other than the starting and ending vertexes.

Although the feasibility of (2.12) with diagonal $P > 0$ can be checked numerically, algebraic conditions like (2.20) that explicitly display the range of feasibility are beneficial when the parameters exhibit wide uncertainty, as in typical biological models. Such conditions further give insight into the interplay between network structure and stability properties.

References

1. Moylan, P., Hill, D.: Stability criteria for large-scale systems. IEEE Trans. Autom. Control **23**(2), 143–149 (1978)
2. Vidyasagar, M.: Input-Output Analysis of Large Scale Interconnected Systems. Springer, Berlin (1981)
3. Grant, M., Boyd, S.: CVX: Matlab software for disciplined convex programming. http://cvxr.com/cvx (2014)
4. Löfberg, J.: Yalmip: a toolbox for modeling and optimization in MATLAB. In: Proceedings of the CACSD Conference, Taipei, Taiwan (2004)
5. Arcak, M., Sontag, E.: Diagonal stability of a class of cyclic systems and its connection with the secant criterion. Automatica **42**(9), 1531–1537 (2006)
6. Arcak, M.: Diagonal stability on cactus graphs and application tonetwork stability analysis. IEEE Trans. Autom. Control **56**(12), 2766–2777 (2011). doi:10.1109/TAC.2011.2125130

Chapter 3
Equilibrium Independent Stability Certification

We consider again the interconnected system (2.1)–(2.3) but now remove the assumption $f_i(0, 0) = 0$, $h_i(0, 0) = 0$ that guaranteed an equilibrium at $x = 0$. We assume an equilibrium

$$x^* = [x_1^{*T} \ldots x_N^{*T}]^T$$

exists, but is not necessarily at the origin. This means that x^* satisfies

$$f_i(x_i^*, u_i^*) = 0 \quad i = 1, \ldots, N \quad \text{where} \quad \underbrace{\begin{bmatrix} u_1^* \\ \vdots \\ u_N^* \end{bmatrix}}_{\triangleq\, u^*} = M \underbrace{\begin{bmatrix} h_1(x_1^*, u_1^*) \\ \vdots \\ h_N(x_N^*, u_N^*) \end{bmatrix}}_{\triangleq\, y^*}. \qquad (3.1)$$

If we can find a storage function $V_i(\cdot)$ for each subsystem such that:

$$V_i(x_i^*) = 0, \quad V_i(x_i) > 0 \quad \forall x_i \neq x_i^*, \quad \text{and}$$

$$\nabla V_i(x_i)^T f_i(x_i, u_i) \leq \begin{bmatrix} u_i - u_i^* \\ y_i - y_i^* \end{bmatrix}^T X_i \begin{bmatrix} u_i - u_i^* \\ y_i - y_i^* \end{bmatrix} \qquad (3.2)$$

then (2.8) with $p_i > 0$ proves stability of x^* as in the previous section.

However, this procedure assumes that x^* is known, which is restrictive. It may be hard to solve the large set of equations (3.1) and, further, the solution depends on the interconnection. Thus, adding or removing subsystems alter x^* and require cumbersome iterations that impair the compositional approach pursued here. Below we define the notion of "equilibrium independent dissipativity" which enables stability certification without the explicit knowledge of x^*.

© The Author(s) 2016
M. Arcak et al., *Networks of Dissipative Systems*,
SpringerBriefs in Control, Automation and Robotics,
DOI 10.1007/978-3-319-29928-0_3

3.1 Equilibrium Independent Dissipativity (EID)

Consider the system

$$\frac{\mathrm{d}}{\mathrm{d}t} x(t) = f(x(t), u(t)) \tag{3.3}$$

$$y(t) = h(x(t), u(t)) \tag{3.4}$$

where $x(t) \in \mathbb{R}^n$, $u(t) \in \mathbb{R}^m$, $y(t) \in \mathbb{R}^p$, and suppose there exists a set $\mathscr{X} \subset \mathbb{R}^n$ where, for every $\bar{x} \in \mathscr{X}$, there is a unique $\bar{u} \in \mathbb{R}^m$ satisfying $f(\bar{x}, \bar{u}) = 0$. Thus $\bar{u}$ and $\bar{y} \triangleq h(\bar{x}, \bar{u})$ are implicit functions of $\bar{x}$.

Definition 3.1 We say that the system above is **equilibrium independent dissipative (EID)** with supply rate $s(\cdot, \cdot)$ if there exists a continuously differentiable storage function $V : \mathbb{R}^n \times \mathscr{X} \mapsto \mathbb{R}$ satisfying, $\forall (x, \bar{x}, u) \in \mathbb{R}^n \times \mathscr{X} \times \mathbb{R}^m$

$$V(x, \bar{x}) \geq 0, \qquad V(\bar{x}, \bar{x}) = 0, \qquad \nabla_x V(x, \bar{x})^T f(x, u) \leq s(u - \bar{u}, y - \bar{y}). \tag{3.5}$$

Unlike (3.2) which is referenced to the equilibrium point x^*, EID demands dissipativity with respect to any point $\bar{x}$ that has the potential to become an equilibrium when the system is interconnected with others. EID was introduced in [1] and refined to the form above in [2]. It was shown in [1] that EID is in general less restrictive than the *incremental dissipativity* notion [3].

For a memoryless system $y(t) = h(u(t))$ we take the storage function to be zero and interpret EID with supply rate $s(\cdot, \cdot)$ as the static inequality

$$s(u - \bar{u}, h(u) - h(\bar{u})) \geq 0 \qquad \forall (u, \bar{u}) \in \mathbb{R}^m \times \mathbb{R}^m. \tag{3.6}$$

As an illustration, for a scalar function $h(\cdot)$ the inequality above with the passivity supply rate $s(u, y) = uy$ is

$$(u - \bar{u})(h(u) - h(\bar{u})) \geq 0 \qquad \forall (u, \bar{u}) \in \mathbb{R} \times \mathbb{R} \tag{3.7}$$

which means that $h(\cdot)$ is increasing[1]:

$$u \geq \bar{u} \implies h(u) \geq h(\bar{u}). \tag{3.8}$$

When $h(\cdot)$ is differentiable (3.7) is equivalent to $h'(u) \geq 0$ for all $u \in \mathbb{R}$. Similarly, (3.6) restricts $h'(u)$ to the interval $[0, 1/\varepsilon]$ for the output strict passivity supply rate $s(u, y) = uy - \varepsilon y^2$, and to $[-\gamma, \gamma]$ for the finite gain supply rate $s(u, y) = \gamma^2 u^2 - y^2$.

---

[1]We refer to (3.8) as an "increasing" property despite the fact that it allows $h(\cdot)$ to be flat. We use the term "strictly increasing" when $u > \bar{u} \implies h(u) > h(\bar{u})$. We follow a similar convention for decreasing functions.

Example 3.1 We examine the equilibrium-independent passivity of

$$\frac{dx(t)}{dt} = f_0(x(t)) + u(t), \quad y(t) = h(x(t)), \quad u(t), x(t), y(t) \in \mathbb{R} \quad (3.9)$$

where $h(\cdot)$ is increasing and $f_0(\cdot)$ is decreasing.

Given $\bar{x} \in \mathbb{R}$, $f(\bar{x}, \bar{u}) = f_0(\bar{x}) + \bar{u} = 0$ admits the unique solution $\bar{u} = -f_0(\bar{x})$. Substituting $f_0(x) + u = f_0(x) - f_0(\bar{x}) + u - \bar{u}$ and $s(u - \bar{u}, y - \bar{y}) = (u - \bar{u})(y - \bar{y}) - \varepsilon(y - \bar{y})^2$, we rewrite (3.5) as

$$\nabla_x V(x, \bar{x})(f_0(x) - f_0(\bar{x}))) + \varepsilon(h(x) - h(\bar{x}))^2 \quad (3.10)$$
$$+ [\nabla_x V(x, \bar{x}) - (h(x) - h(\bar{x}))](u - \bar{u}) \le 0.$$

Thus, we seek a $V(\cdot, \cdot)$ such that $V(x, \bar{x}) \ge 0$, $V(\bar{x}, \bar{x}) = 0$ for all x, $\bar{x}$, and (3.10) holds with $\varepsilon \ge 0$.

Note that (3.10) implies

$$\nabla_x V(x, \bar{x}) = h(x) - h(\bar{x}) \quad (3.11)$$

because, if $\nabla_x V(x, \bar{x}) - (h(x) - h(\bar{x})) \ne 0$ for some x, we can select u such that $[\nabla_x V(x, \bar{x}) - (h(x) - h(\bar{x}))](u - \bar{u})$ is positive and large enough to contradict (3.10). To satisfy (3.11) as well as $V(\bar{x}, \bar{x}) = 0$ we let

$$V(x, \bar{x}) = \int_{\bar{x}}^{x} [h(z) - h(\bar{x})]dz \quad (3.12)$$

which further satisfies $V(x, \bar{x}) \ge 0$ because $h(\cdot)$ is increasing. Thus (3.10) becomes

$$(h(x) - h(\bar{x}))[(f_0(x) + \varepsilon h(x)) - (f_0(\bar{x}) + \varepsilon h(\bar{x}))] \le 0. \quad (3.13)$$

For $\varepsilon = 0$ this inequality follows from the decreasing property of $f_0(\cdot)$, because the sign of $(h(x) - h(\bar{x}))$ is the same as $(x - \bar{x})$ and the sign of $(f_0(x) - f_0(\bar{x}))$ is opposite to $(x - \bar{x})$. Thus we conclude equilibrium independent passivity without further assumptions.

If, in addition, $f_0(\cdot) + \varepsilon h(\cdot)$ remains decreasing up to some $\varepsilon > 0$, then a similar sign argument guarantees (3.13), proving equilibrium-independent output strict passivity.

We next generalize the model (3.9) to

$$\frac{dx(t)}{dt} = f_0(x(t)) + g(x(t))u(t), \quad y(t) = h(x(t)), \quad u(t), x(t), y(t) \in \mathbb{R} \quad (3.14)$$

which contains the new function $g(\cdot)$, assumed to satisfy $g(x) > 0$ for all x. With the modified storage function

$$V(x, \bar{x}) = \int_{\bar{x}}^{x} \frac{h(z) - h(\bar{x})}{g(z)} dz \tag{3.15}$$

we get

$$\nabla_x V(x, \bar{x})(f_0(x) + g(x)u) = (h(x) - h(\bar{x})) \left(\frac{f_0(x)}{g(x)} + u \right)$$

$$= (h(x) - h(\bar{x})) \left(\frac{f_0(x)}{g(x)} - \frac{f_0(\bar{x})}{g(\bar{x})} + u - \bar{u} \right).$$

Arguments similar to those for $g(x) \equiv 1$ above yield the following conclusion:

The system (3.14) is equilibrium independent passive if $g(x) > 0$ for all x, $h(\cdot)$ is increasing, and

$$\theta(\cdot) \triangleq \frac{f_0(\cdot)}{g(\cdot)} \tag{3.16}$$

is decreasing. It is equilibrium independent *output strictly* passive if

$$\theta(\cdot) + \varepsilon h(\cdot)$$

remains decreasing up to some $\varepsilon > 0$.

3.2 Numerical Certification of EID

For linear systems, EID coincides with standard dissipativity. To see this let

$$f(x, u) = Ax + Bu \quad h(x, u) = Cx + Du$$

and note that if B is full column rank then there exists unique $\bar{u}$ satisfying

$$A\bar{x} + B\bar{u} = 0$$

when $\bar{x}$ is constrained to an appropriate subspace. Substituting $f(x, u) = A(x - \bar{x}) + B(u - \bar{u})$ and the candidate storage function

$$V(x, \bar{x}) = \frac{1}{2}(x - \bar{x})^T P(x - \bar{x})$$

in (3.5) we get the EID condition

$$(x - \bar{x})^T P[A(x - \bar{x}) + B(u - \bar{u})] \leq s(u - \bar{u}, C(x - \bar{x}) + D(u - \bar{u}))$$

which is identical to standard dissipativity, with shifted variables.

For polynomial systems, certifying EID can be cast as a SOS feasibility program. Recall that we denote the set of all polynomials in x as $\mathbb{R}[x]$ and all SOS polynomials as $\Sigma[x]$. A polynomial system is EID with respect to a polynomial supply rate s if there exists functions V and r satisfying

$$
\begin{aligned}
V(x, \bar{x}) &\in \Sigma[x, \bar{x}] \\
r(x, u, \bar{x}, \bar{u}) &\in \mathbb{R}[x, u, \bar{x}, \bar{u}] \\
-\nabla_x V(x, \bar{x})^T f(x, u) + s(u - \bar{u}, h(x, u) - h(\bar{x}, \bar{u})) & \\
+ r(x, u, \bar{x}, \bar{u}) f(\bar{x}, \bar{u}) &\in \Sigma[x, u, \bar{x}, \bar{u}].
\end{aligned}
\tag{3.17}
$$

The constraint $V(\bar{x}, \bar{x}) = 0$ is enforced by letting $V(x, \bar{x}) = (x - \bar{x})^T Q(x, \bar{x})(x - \bar{x})$ where $Q(x, \bar{x})$ is a symmetric matrix of polynomials.

Note that $\bar{x}$ and $\bar{u}$ are variables and not assumed to satisfy $f(\bar{x}, \bar{u}) = 0$. Instead, the term $r(x, u, \bar{x}, \bar{u}) f(\bar{x}, \bar{u})$ ensures that whenever $f(\bar{x}, \bar{u}) = 0$ then

$$\nabla_x V(x, \bar{x})^T f(x, u) \leq s(u - \bar{u}, h(x, u) - h(\bar{x}, \bar{u}))$$

for all $x \in \mathbb{R}^n$, $u \in \mathbb{R}^m$ as desired.

3.3 The Stability Theorem

We return to the interconnected system (2.1)–(2.3) and assume that an equilibrium x^* exists as in (3.1). With the notion of EID we no longer rely on the explicit knowledge of x^* to certify stability.

Theorem 3.1 *Suppose the interconnected system (2.1)–(2.3) admits an equilibrium x^* as in (3.1) and each subsystem is EID with a quadratic supply rate (2.4) and storage function $V_i(\cdot, \cdot)$ satisfying $V_i(\bar{x}, \bar{x}) = 0$, and $V_i(x_i, \bar{x}) > 0$ when $x_i \neq \bar{x}$. If there exist $p_i > 0$, $i = 1, \ldots, N$, such that (2.8) holds, then x^* is stable and a Lyapunov function is*

$$V(x) = p_1 V_1(x_1, x_1^*) + \cdots + p_N V_N(x_N, x_N^*).$$

This expression defines a family of Lyapunov functions parameterized by the weights p_i and the equilibrium x^*. However, to infer stability we need neither the weights nor the equilibrium explicitly.

References

1. Hines, G., Arcak, M., Packard, A.: Equilibrium-independent passivity: a new definition and numerical certification. Automatica **47**(9), 1949–1956 (2011)
2. Bürger, M., Zelazo, D., Allgöwer, F.: Duality and network theory in passivity-based cooperative control. Automatica **50**(8), 2051–2061 (2014)
3. Stan, G.B., Sepulchre, R.: Analysis of interconnected oscillators by dissipativity theory. IEEE Trans. Autom. Control **52**(2), 256–270 (2007)

Chapter 4
Case Studies

4.1 A Cyclic Biochemical Reaction Network

Consider the following model of a mitogen-activated protein kinase (MAPK) cascade with inhibitory feedback:

$$\frac{dx_1(t)}{dt} = -\frac{b_1 x_1(t)}{c_1 + x_1(t)} + \frac{d_1(1 - x_1(t))}{e_1 + (1 - x_1(t))} \frac{\mu}{1 + kx_3(t)}$$
$$\frac{dx_2(t)}{dt} = -\frac{b_2 x_2(t)}{c_2 + x_2(t)} + \frac{d_2(1 - x_2(t))}{e_2 + (1 - x_2(t))} x_1(t) \qquad (4.1)$$
$$\frac{dx_3(t)}{dt} = -\frac{b_3 x_3(t)}{c_3 + x_3(t)} + \frac{d_3(1 - x_3(t))}{e_3 + (1 - x_3(t))} x_2(t).$$

The variable $x_i \in [0, 1]$, $i = 1, 2, 3$, denotes the concentration of the phosphorylated (active) form of the protein M_i in Fig. 4.1, and $1 - x_i$ is the concentration of the inactive form (after a suitable scaling that brings the total concentration to one). All parameters are positive.

The second term in each equation is the rate of activation and the first term is the rate of inactivation for the respective protein. For $i = 2, 3$ the activation rate is proportional to x_{i-1}, which means that the phosphorylated protein upstream facilitates downstream activation. In contrast, the activation of M_1 is inhibited by the active form of M_3, as represented by the decreasing function $\mu/(1 + kx_3)$ and depicted with a dashed line in Fig. 4.1.

The inhibition of the first stage of the cascade by the last stage is a feedback regulation, comparable to an assembly line where the most upstream workstation is decelerated when the final product starts piling up at the end of the line.

A strong negative feedback of this form may generate oscillations which, for a MAPK cascade, means a transient response to a stimulus rather than sustained activation. Temporal patterns of activation are believed to determine cell fate [1]

© The Author(s) 2016
M. Arcak et al., *Networks of Dissipative Systems*,
SpringerBriefs in Control, Automation and Robotics,
DOI 10.1007/978-3-319-29928-0_4

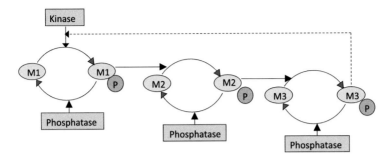

Fig. 4.1 A mitogen-activated protein kinase (MAPK) cascade with inhibitory feedback. *Solid lines* represent activation; the *dashed line* represents inhibition

(e.g., proliferation in response to transient activation versus differentiation in response to sustained activation), thus motivating dynamical analysis.

We decompose the system (4.1) as in the negative feedback cyclic interconnection of Fig. 2.3, where the subsystems are

$$G_i : \quad \frac{dx_i(t)}{dt} = f_i(x_i(t)) + g_i(x_i(t))u_i(t) \quad y_i(t) = h_i(x_i(t)) \qquad (4.2)$$

$i = 1, 2, 3$, and the functions $f_i(\cdot)$, $g_i(\cdot)$, $h_i(\cdot)$ are defined as

$$f_i(x_i) = -\frac{b_i x_i}{c_i + x_i} \quad g_i(x_i) = \frac{d_i(1 - x_i)}{e_i + (1 - x_i)} \quad i = 1, 2, 3$$

$$h_i(x_i) = x_i \quad i = 1, 2, \quad h_3(x_3) = -\frac{\mu}{1 + kx_3}.$$

Each subsystem is of the form (3.14) studied in Example 3.1 where $h_i(\cdot)$ is increasing and $\theta_i(\cdot)$ defined by

$$\theta_i(x) = \frac{f_i(x)}{g_i(x)} \qquad (4.3)$$

is decreasing. Thus, we estimate the largest $\varepsilon_i > 0$ such that $\theta_i(\cdot) + \varepsilon_i h_i(\cdot)$ is decreasing and apply the stability criterion (2.14) for cyclic interconnections.

To show that a steady state x^* exists we first note that each $\theta_i : [0, 1] \mapsto (-\infty, 0]$ is strictly decreasing and onto; therefore, $\theta_i^{-1} : (-\infty, 0] \mapsto [0, 1]$ is well-defined and decreasing. Next, note that the steady state equations

$$\theta_i(x_i^*) + u_i^* = 0 \quad i = 1, 2, 3, \quad u_2^* = x_1^*, \ u_3^* = x_2^*, \ u_1^* = -h_3(x_3^*)$$

imply

$$\theta_1(x_1^*) = h_3(\theta_3^{-1}(-\theta_2^{-1}(-x_1^*)))$$

where the left-hand side is the strictly decreasing and onto function $\theta_1 : [0, 1] \mapsto (-\infty, 0]$ and the right-hand side is an increasing function with negative values. Thus, the two functions intersect at a unique point x_1^*. This implies that a steady state x^* exists and is unique.

If ε_i, $i = 1, 2, 3$, satisfy (2.14), then the stability of x^* is ascertained with a Lyapunov function that is a weighted sum of storage functions of the form (3.15):

$$V(x) = p_1 \int_{x_1^*}^{x_1} \frac{z - x_1^*}{g_1(z)} dz + p_2 \int_{x_2^*}^{x_2} \frac{z - x_2^*}{g_2(z)} dz + p_3 \int_{x_3^*}^{x_3} \frac{h_3(z) - h_3(x_3^*)}{g_3(z)} dz.$$

The weights $p_i > 0$ are obtained from the LMI (2.12) which is guaranteed to have a diagonal solution $P > 0$ by (2.14).

Note from the explicit form of the functions $g_i(\cdot)$ and $h_3(\cdot)$ that $V(\cdot)$ above is not an apparent choice for a Lyapunov function. It further depends on the implicit solution for x^* whose existence and uniqueness were argued only qualitatively.

For the numerical details of estimating the parameters ε_i, $i = 1, 2, 3$, such that $\theta_i(\cdot) + \varepsilon_i h_i(\cdot)$ is decreasing, we refer the reader to [2]. Other feedback structures of MAPK cascades were also studied in [2] with the approach illustrated in this example.

4.2 A Vehicle Platoon

Consider a platoon where the velocity of each vehicle is governed by

$$\frac{dv_i(t)}{dt} = -v_i(t) + v_i^0 + u_i(t) \qquad i = 1, \ldots, N \tag{4.4}$$

in which $u_i(t)$ is a coordination feedback to be designed, and v_i^0 is the nominal velocity of vehicle i in the absence of feedback. The position of vehicle i is then obtained from

$$\frac{dx_i(t)}{dt} = v_i(t).$$

We will design feedback laws that depend on relative positions with respect to a subset of other vehicles, typically nearest neighbors.

We introduce an undirected graph where the vertices represent the vehicles and an edge between vertices i and j means that vehicles i and j have access to the relative position measurement $x_i(t) - x_j(t)$. Next we assign an orientation to each edge by selecting one end to be the head and the other to be the tail. Then the *incidence matrix*

$$D_{il} = \begin{cases} 1 & \text{if vertex } i \text{ is the head of edge } l \\ -1 & \text{if vertex } i \text{ is the tail of edge } l \\ 0 & \text{otherwise} \end{cases} \tag{4.5}$$

Fig. 4.2 A vehicle platoon. The motion of the vehicles is coordinated with relative position feedback

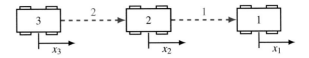

generates a vector of relative positions z_l for the edges $l = 1, \ldots, L$ by

$$z = D^T x. \tag{4.6}$$

As an illustration, in Fig. 4.2,

$$D = \begin{bmatrix} 1 & 0 \\ -1 & 1 \\ 0 & -1 \end{bmatrix} \quad \text{and} \quad \begin{bmatrix} z_1 \\ z_2 \end{bmatrix} = D^T x = \begin{bmatrix} x_1 - x_2 \\ x_2 - x_3 \end{bmatrix}.$$

We propose the feedback law

$$u = -D \begin{bmatrix} h_1(z_1) \\ \vdots \\ h_L(z_L) \end{bmatrix} \tag{4.7}$$

where each function $h_l : \mathbb{R} \mapsto \mathbb{R}$ is strictly increasing and onto. This means that vehicle i applies the input

$$u_i = -\sum_{l=1}^{L} D_{il} h_l(z_l) \tag{4.8}$$

which depends on locally available measurements because $D_{il} \neq 0$ only when vertex i is the head or tail of edge l. In the case of Fig. 4.2,

$$u_1 = -h_1(z_1) \quad u_2 = h_1(z_1) - h_2(z_2) \quad u_3 = h_2(z_2)$$

where we may interpret $h_1(z_1)$ and $h_2(z_2)$ as virtual spring forces between vehicles 1 and 2, and 2 and 3, respectively.

We now analyze the stability of the equilibrium whose existence and uniqueness will be discussed subsequently. We note from (4.6) that

$$\frac{\mathrm{d}}{\mathrm{d}t} z(t) = D^T v(t) \triangleq w(t) \tag{4.9}$$

where we interpret $w(t)$ as an input and define the output

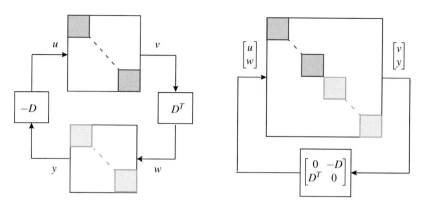

Fig. 4.3 A block diagram for the platoon dynamics. *Left*: the feedforward blocks $u_i \mapsto v_i$ represent the velocity dynamics (4.4) and the feedback blocks $w_l \mapsto y_l$ represent the lth subsystem of the relative position dynamics (4.9)–(4.10). *Right*: the diagram on the *left* brought to the canonical form of Fig. 2.1

$$y(t) \triangleq \begin{bmatrix} h_1(z_1(t)) \\ \vdots \\ h_L(z_L(t)) \end{bmatrix}. \tag{4.10}$$

We then represent the closed-loop system with the block diagram of Fig. 4.3 (left) where the feedforward blocks $u_i \mapsto v_i$ represent the velocity dynamics (4.4) and the feedback blocks $w_l \mapsto y_l$ represent the lth subsystem of the relative position dynamics (4.9)–(4.10). This block diagram is equivalent to the one on the right which is of the standard form in Fig. 2.1 with the interconnection matrix

$$M = \begin{bmatrix} 0 & -D \\ D^T & 0 \end{bmatrix}.$$

Noting that M is skew symmetric as in Sect. 2.3.1 we proceed to proving that each subsystem is equilibrium-independent passive. To do so we compare each to (3.9) in Example 3.1 which we found to be equilibrium-independent passive when $f_0(\cdot)$ is decreasing and $h(\cdot)$ is increasing. This is indeed the case for the $w_l \mapsto y_l$ subsystems in (4.9)–(4.10) where $f_0(z_l) = 0$. The $u_i \mapsto v_i$ subsystems in (4.4), where $f_0(v_i) = -v_i + v_i^0$, $h(v_i) = v_i$ are equilibrium-independent *output strictly passive* because $f_0(\cdot) + \varepsilon h(\cdot)$ remains decreasing up to $\varepsilon = 1$.

We thus conclude that if an equilibrium

$$z_l = z_l^*, \ l = 1, \ldots, L, \quad v_i = v_i^*, \ i = 1, \ldots, N,$$

exists, it is stable from the equilibrium-independent passivity of the subsystems and the skew symmetry of their interconnection.

At equilibrium the right-hand side of (4.9) must vanish, that is

$$D^T v^* = 0. \tag{4.11}$$

By the definition (4.5) above, the null space of D^T includes the vector of ones: $D^T \mathbf{1} = 0$. In addition, if the graph is connected then the span of $\mathbf{1}$ constitutes the entire null space: there is no solution to (4.11) other than $v^* = \vartheta \mathbf{1}$ where ϑ is a common platoon velocity.

Setting the right-hand side of (4.4) to zero, we see that the equilibrium value of the inputs u_i must compensate for the variations in the nominal velocities v_i^0 so that a common velocity ϑ can be maintained:

$$-\vartheta + v_i^0 + u_i^* = 0 \quad i = 1, \ldots, N. \tag{4.12}$$

Note that $\sum_{i=1}^N u_i = \mathbf{1}^T u = 0$, which follows from (4.7) and $\mathbf{1}^T D = 0$. Thus, adding the equations (4.12) from $i = 1$ to $i = N$ we get

$$-N\vartheta + \sum_{i=1}^N v_i^0 = 0$$

which shows that the common velocity ϑ must be the average $\frac{1}{N}\sum_{i=1}^N v_i^0$.

Substituting this average for ϑ and (4.8) for u_i^* back in (4.12), we obtain the following equations for z_l^*:

$$v_i^0 - \frac{1}{N}\sum_{i=1}^N v_i^0 = \sum_{l=1}^L D_{il} h_l(z_l^*) \quad i = 1, \ldots, N.$$

These equations are particularly transparent for a line graph as in Fig. 4.2 where the head and tail of edge l are vertices l and $l + 1$:

$$v_1^0 - \frac{1}{N}\sum_{i=1}^N v_i^0 = h_1(z_1^*)$$

$$v_i^0 - \frac{1}{N}\sum_{i=1}^N v_i^0 = -h_{i-1}(z_{i-1}^*) + h_i(z_i^*) \quad i = 2, \ldots, N-1$$

$$v_N^0 - \frac{1}{N}\sum_{i=1}^N v_i^0 = -h_{N-1}(z_{N-1}^*).$$

Adding equations $i = 1$ to l above we get a new equation that depends only on $h_l(z_l^*)$. Then a solution z_l^* exists since $h_l(\cdot)$ is onto, and is unique since $h_l(\cdot)$ is strictly increasing. A similar argument may be developed for other acyclic graphs.

The proof is more elaborate for graphs with cycles where the variables z_l are now interdependent through algebraic constraints [3].

4.3 Internet Congestion Control

The congestion control problem is to maximize the network throughput while ensuring an equitable allocation of bandwidth to the users. In a decentralized congestion control scheme each link increases its packet drop or marking probability, interpreted as the "price" of the link, as the transmission rate approaches the capacity of the link. Sources then adjust their sending rates based on the aggregate price feedback they receive in the form of dropped or marked packets.

To see the interconnection structure of sources and links, consider a network where packets from sources $i = 1, \ldots, N$ are routed through links $j = 1, \ldots, L$ according to a $L \times N$ routing matrix

$$R_{li} = \begin{cases} 1 & \text{if source } i \text{ uses link } l \\ 0 & \text{otherwise.} \end{cases} \tag{4.13}$$

An example with $N = 3$ sources and $L = 2$ links is shown in Fig. 4.4.

Because the transmission rate w_j of link j is the sum of the sending rates v_i of sources using that link, the vectors of link rates w and source rates v are related by

$$w = Rv. \tag{4.14}$$

Likewise, the total price feedback q_i received by source i is the sum of the prices p_j of the links on its path, which implies

$$q = R^T p. \tag{4.15}$$

Representing the user algorithms as subsystems $G_i : -q_i \mapsto v_i, i = 1, \ldots, N$ and the router algorithms as $G_{N+j} : w_j \mapsto p_j, j = 1, \ldots, L$, we get an interconnection as in the standard form of Fig. 2.1 with

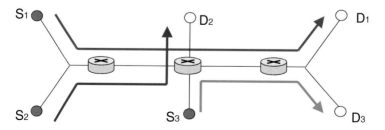

Fig. 4.4 A network with $N = 3$ sources and $L = 2$ links. The rows of the 2×3 routing matrix are [1 1 0] and [1 0 1] corresponding, respectively, to the links on the left and right

$$M = \begin{bmatrix} 0 & -R^T \\ R & 0 \end{bmatrix}. \qquad (4.16)$$

This interconnection is skew symmetric and has the same structure as Fig. 4.3 of the platoon example, with the routing matrix R replacing D^T, the feedforward blocks now representing user algorithms, and the feedback blocks representing router algorithms. Thus, by imposing passivity as a requirement for these algorithms, we guarantee network stability without further restrictions.

As an illustration, in Kelly's primal algorithm [4] the user update is

$$G_i : \quad \frac{d}{dt} v_i(t) = g_i(v_i(t))(U_i'(v_i(t)) - q_i(t)) \quad i = 1, \ldots, N \qquad (4.17)$$

where $g_i(v_i) > 0$, for all $v_i \geq 0$ and $U_i'(\cdot)$ is the derivative of a concave utility function $U_i : \mathbb{R}_{\geq 0} \mapsto \mathbb{R}$ where we further assume

$$U_i'(v_i) \to \infty \quad \text{as} \quad v_i \to 0^+. \qquad (4.18)$$

The router update is

$$G_{j+N} : \quad p_j(t) = h_j(w_j(t)) \quad j = 1, \ldots, L \qquad (4.19)$$

where $h_j : \mathbb{R}_{\geq 0} \mapsto \mathbb{R}_{\geq 0}$ is an increasing function.

Condition (4.18) enforces the physical constraint $v_i \geq 0$ for the solutions of (4.17), and mild additional assumptions[1] guarantee a unique equilibrium in $\mathbb{R}_{\geq 0}^N$. This equilbrium approximates the Kuhn–Tucker optimality conditions for the problem

$$\max_{v_i \geq 0} \sum_i U_i(v_i) \quad \text{subject to} \quad w_j \leq c_j$$

when $h_j(\cdot)$ is interpreted as a penalty function that increases with a steep slope as w_j approaches the link capacity c_j.

To ascertain the stability of this equilibrium without relying on its explicit knowledge, we proceed to analyze the equilibrium-independent passivity properties of the subsystems above.

The router algorithm (4.19) is static and, thus, equilibrium-independent passivity follows from the increasing property of $h_j(\cdot)$. The user algorithm (4.17) is of the form (3.14) in Example 3.1 with input $u_i = -q_i$ and output v_i. The function $U_i'(\cdot)$ plays the role of $\theta(\cdot)$ in (3.16) and is decreasing thanks to the concavity of $U_i(\cdot)$. Thus, the storage function

$$V_i(v_i, \bar{v}_i) = \int_{\bar{v}_i}^{v_i} \frac{z - \bar{v}_i}{g_i(z)} dz \qquad (4.20)$$

---

[1]For example, the strict concavity condition (4.21) is sufficient for the existence of a unique equilibrium [5].

guarantees equilibrium-independent passivity. If, in addition,

$$U_i''(v_i) \leq -\varepsilon < 0 \tag{4.21}$$

for all $v_i \geq 0$, then $U_i'(v_i) + \varepsilon v_i$ is a decreasing function of v_i and we conclude equilibrium-independent *output strict* passivity.

Since the interconnection is skew symmetric, the stability criterion (2.12) holds with $P = I$ and the sum of the storage functions (4.20) serves as a Lyapunov function. Similar Lyapunov constructions from storage functions were pursued for Kelly's dual algorithm in [5] and for a primal-dual algorithm in [6].

4.4 Population Dynamics of Interacting Species

Consider the following model for N interacting species:

$$\frac{d}{dt}x_i(t) = \left(\lambda_i - \gamma_i x_i(t) + \sum_{j \neq i} m_{ij} x_j(t) \right) x_i(t) \quad i = 1, 2, \ldots, N \tag{4.22}$$

where x_i is the population of species i, and λ_i and γ_i are positive coefficients.

When $N = 1$, we recover the *logistic growth* model [7] which admits a stable equilibrium at the carrying capacity $x_i = \lambda_i/\gamma_i$. When $N = 2$, (4.22) encompasses models of mutualism ($m_{12} > 0$, $m_{21} > 0$), competition ($m_{12} < 0$, $m_{21} < 0$), and predation ($m_{12}m_{21} < 0$).

We decompose (4.22) into the subsystems

$$G_i : \quad \frac{d}{dt}x_i(t) = (\lambda_i - \gamma_i x_i(t))x_i(t) + x_i(t)u_i(t) \quad y_i(t) = x_i(t) \quad i = 1, 2, \ldots, N, \tag{4.23}$$

interconnected as in Fig. 2.1, where $M = (m_{ij}) \in \mathbb{R}^{N \times N}$ with diagonal entries m_{ii} interpreted as zero.

Note the each G_i is of the form (3.14) with $g(x) = h(x) = x$, and $\theta(x) = \lambda_i - \gamma_i x$ as defined in (3.16). Since $\theta(x) + \varepsilon_i x$ is a decreasing function of x up to $\varepsilon_i = \gamma_i^{-1}$, we conclude equilibrium-independent output strict passivity, and the storage function in (3.15) takes the form

$$V_i(x_i, \bar{x}_i) = x_i - \bar{x}_i - \bar{x}_i \ln\left(\frac{x_i}{\bar{x}_i}\right). \tag{4.24}$$

Thus, if an equilibrium x^* exists in the positive orthant and if (2.12) with

$$E = \mathrm{diag}(\gamma_1^{-1}, \ldots, \gamma_N^{-1})$$

admits a diagonal solution $P > 0$, the stability of x^* is certified with the Lyapunov function

$$V(x) = \sum_{i=1}^{N} p_i \left\{ x_i - x_i^* - x_i^* \ln \left(\frac{x_i}{x_i^*} \right) \right\}. \tag{4.25}$$

Asymptotic stability follows when (2.12) holds with strict inequality.

Two Species

When $N = 2$ and $m_{12}m_{21} < 0$ (predation) the incidence graph of M consists of a single cycle with negative gain and length two (Sect. 2.3.3). This means that $\alpha = 0$ in (2.20), and (2.12) is strictly feasible with diagonal $P > 0$. Thus, the equilibrium x^* is asymptotically stable.

When $m_{12}m_{21} > 0$ (mutualism or competition) the cycle gain is positive and, by (2.20), feasibility is equivalent to

$$\Gamma = \frac{m_{12}m_{21}}{\varepsilon_1\varepsilon_2} = m_{12}m_{21}\gamma_1\gamma_2 < 1.$$

Antelopes, Hyenas, and Lions

As another example suppose species 2 and 3 both prey on species 1:

$$m_{12} < 0 \quad m_{13} < 0 \quad m_{21} > 0 \quad m_{31} > 0, \tag{4.26}$$

but are neutral to each other

$$m_{23} = m_{32} = 0. \tag{4.27}$$

This means that the incidence graph of M consists of two cycles that intersect at vertex 1 as in Fig. 4.5, thus conforming to the cactus structure described in Sect. 2.3.3. Each cycle has negative gain and length two, therefore $\alpha_1 = \alpha_2 = 0$ in (2.20), and (2.12)

Fig. 4.5 The incidence graph of matrix M with sign structure (4.26)–(4.27)

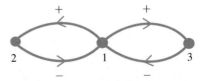

is strictly feasible with a diagonal $P > 0$. Thus, the equilibrium x^* is asymptotically stable without restrictions on the model parameters other than the sign conditions (4.26)–(4.27).

References

1. Kholodenko, B.: Cell-signalling dynamics in time and space. Nat. Rev. Mol. Cell Biol. **7**, 165–176 (2006)
2. Arcak, M., Sontag, E.: A passivity-based stability criterion for a class of biochemical reaction networks. Math. Biosci. Eng. **5**(1), 1–19 (2008)
3. Bürger, M., Zelazo, D., Allgöwer, F.: Duality and network theory in passivity-based cooperative control. Automatica **50**(8), 2051–2061 (2014)
4. Kelly, F., Maulloo, A., Tan, D.: Rate control in communication networks: shadow prices, proportional fairness and stability. J. Oper. Res. Soc. **49**, 237–252 (1998)
5. Wen, J., Arcak, M.: A unifying passivity framework for network flow control. IEEE Trans. Autom. Control **49**(2), 162–174 (2004)
6. Srikant, R.: The Mathematics of Internet Congestion Control. Birkhauser, Boston (2004)
7. Murray, J.: Mathematical Biology, I: an Introduction, 3rd edn. Springer, New York (2001)

Chapter 5
From Stability to Performance and Safety

5.1 Compositional Performance Certification

Consider the interconnection in Fig. 5.1, modified from Fig. 2.1 to accommodate an exogenous disturbance input $d \in \mathbb{R}^m$ and to define a performance output $e \in \mathbb{R}^p$. The matrix $\overline{M}$ specifies the subsystem inputs and the performance output by

$$\begin{bmatrix} u \\ e \end{bmatrix} = \overline{M} \begin{bmatrix} y \\ d \end{bmatrix} = \begin{bmatrix} M_{uy} & M_{ud} \\ M_{ey} & M_{ed} \end{bmatrix} \begin{bmatrix} y \\ d \end{bmatrix} \tag{5.1}$$

where the upper left block M_{uy}, mapping y to u, plays the role of M in Fig. 2.1.

The goal is now to certify the dissipativity of the interconnected system with respect to the supply rate

$$\begin{bmatrix} d \\ e \end{bmatrix}^T W \begin{bmatrix} d \\ e \end{bmatrix} \tag{5.2}$$

where the choice of W signifies a performance objective, such as a prescribed L_2 gain from the disturbance to the performance output. To reach this goal we employ the candidate storage function

$$V(x) = p_1 V_1(x_1) + \cdots + p_N V_N(x_N), \tag{5.3}$$

$p_i \geq 0, i = 1, \ldots, N$, and recall that it satisfies (2.6). The right-hand side of (2.6), rewritten as in (2.7), is indeed dominated by the performance supply rate (5.2) if

$$\begin{bmatrix} u \\ y \\ d \\ e \end{bmatrix}^T \begin{bmatrix} \mathbf{X}(p_1 X_1, \ldots, p_N X_N) & 0 \\ 0 & -W \end{bmatrix} \begin{bmatrix} u \\ y \\ d \\ e \end{bmatrix} \leq 0 \tag{5.4}$$

© The Author(s) 2016
M. Arcak et al., *Networks of Dissipative Systems*,
SpringerBriefs in Control, Automation and Robotics,
DOI 10.1007/978-3-319-29928-0_5

Fig. 5.1 Interconnected
system with exogenous input
d and performance output e

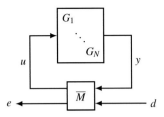

when the variables u and e are constrained by (5.1). Substituting

$$
\begin{bmatrix} u \\ y \\ d \\ e \end{bmatrix} = \begin{bmatrix} M_{uy} & M_{ud} \\ I & 0 \\ 0 & I \\ M_{ey} & M_{ed} \end{bmatrix} \begin{bmatrix} y \\ d \end{bmatrix}
\tag{5.5}
$$

in (5.4), we obtain the performance condition (5.6) below.

Proposition 5.1 *Suppose each subsystem G_i, defined in (2.1) and (2.2) with $f_i(0, 0) = 0$, $h_i(0, 0) = 0$, is dissipative with the quadratic supply rate (2.4) and storage function $V_i(\cdot)$ such that $V_i(0) = 0$ and $V_i(x_i) \geq 0 \; \forall x_i$. If there exist $p_i \geq 0$, $i = 1, \ldots, N$, such that*

$$
\begin{bmatrix} M_{uy} & M_{ud} \\ I & 0 \\ 0 & I \\ M_{ey} & M_{ed} \end{bmatrix}^T \begin{bmatrix} \mathbf{X}(p_1 X_1, \ldots, p_N X_N) & 0 \\ 0 & -W \end{bmatrix} \begin{bmatrix} M_{uy} & M_{ud} \\ I & 0 \\ 0 & I \\ M_{ey} & M_{ed} \end{bmatrix} \leq 0 \quad (5.6)
$$

where $\mathbf{X}(p_1 X_1, \ldots, p_N X_N)$ is as defined in (2.7), then the interconnection is dissipative with respect to the supply rate (5.2), and (5.3) is a storage function.

Note that the stability condition (2.8) is a special case of (5.6) with $W = 0$ where M in (2.8) corresponds to M_{uy} in (5.6). However, when applying the stability condition (2.8) we require positive definiteness of the storage functions $V_i(\cdot)$ and strict positivity of the weights p_i.

We next describe the modifications needed when the assumption $f_i(0, 0) = 0$, $h_i(0, 0) = 0$ is removed from the proposition above. Instead, we assume an equilibrium x^*, whose numerical value is not explicitly known, exists as in (3.1) with $M = M_{uy}$. We wish to establish dissipativity with respect to the supply rate

$$
\begin{bmatrix} d \\ e - e^* \end{bmatrix}^T W \begin{bmatrix} d \\ e - e^* \end{bmatrix}
\tag{5.7}
$$

which depends on the deviation of the performance output e from its equilibrium value $e^* = M_{ey} y^*$.

If each subsystem is EID as in (3.5) with a quadratic supply rate (2.4), then

$$V(x) = p_1 V_1(x_1, x_1^*) + \cdots + p_N V_N(x_N, x_N^*),$$ (5.8)

$p_i \geq 0, i = 1, \ldots, N$, satisfies

$$\sum_{i=1}^{N} p_i \nabla_{x_i} V_i(x_i, x_i^*)^T f_i(x_i, u_i)$$

$$\leq \begin{bmatrix} u - u^* \\ y - y^* \end{bmatrix}^T \mathbf{X}(p_1 X_1, \ldots, p_N X_N) \begin{bmatrix} u - u^* \\ y - y^* \end{bmatrix}.$$ (5.9)

Since $u^* = M_{uy} y^*$ and $e^* = M_{ey} y^*$, it follows from (5.1) that

$$\begin{bmatrix} u - u^* \\ e - e^* \end{bmatrix} = \begin{bmatrix} M_{uy} & M_{ud} \\ M_{ey} & M_{ed} \end{bmatrix} \begin{bmatrix} y - y^* \\ d \end{bmatrix}.$$ (5.10)

Thus, (5.6) implies

$$\begin{bmatrix} u - u^* \\ y - y^* \\ d \\ e - e^* \end{bmatrix}^T \begin{bmatrix} \mathbf{X}(p_1 X_1, \ldots, p_N X_N) & 0 \\ 0 & -W \end{bmatrix} \begin{bmatrix} u - u^* \\ y - y^* \\ d \\ e - e^* \end{bmatrix} \leq 0$$ (5.11)

which allows us to upper bound the right-hand side of (5.9) with (5.7).

We conclude that Proposition 5.1 above holds with the supply rate (5.7) if we remove the restriction $f_i(0, 0) = 0, h_i(0, 0) = 0$, instead strengthening the subsystem dissipativity assumption with its equilibrium independent form.

5.2 Safety under Finite Energy Inputs

In this section we assume the disturbance in Fig. 5.1 has finite L_2 norm,

$$\|d\|_2^2 = \int_0^\infty |d(t)|^2 dt \leq \beta,$$ (5.12)

and aim to certify the following *safety* property for the interconnection:

Trajectories starting from $x(0) = 0$ cannot intersect a given unsafe set $\mathcal{U}$ for any disturbance $d(\cdot)$ satisfying (5.12).

To achieve this goal, we employ the L_2 reachability supply rate $s(d, e) = |d|^2$ from Sect. 1.6, that is (5.2) with

$$W = \begin{bmatrix} I_m & 0 \\ 0 & 0 \end{bmatrix}. \tag{5.13}$$

If (5.6) holds with this W then

$$V(x) = p_1 V_1(x_1) + \cdots + p_N V_N(x_N)$$

satisfies $V(x(\tau)) \le \|d\|_2^2$ for all $\tau \ge 0$. To certify safety for all $d(\cdot)$ with $\|d\|_2^2 \le \beta$, the task is to guarantee that the sublevel set

$$\mathscr{V}_\beta \triangleq \{x : V(x) \le \beta\}$$

does not intersect the unsafe set $\mathscr{U}$; that is, its complement $\overline{\mathscr{V}}_\beta$ contains $\mathscr{U}$:

$$\mathscr{U} \subset \overline{\mathscr{V}}_\beta. \tag{5.14}$$

To apply SOS techniques to this task, assume each V_i is a polynomial and that $\mathscr{U}$ is defined as

$$\mathscr{U} \triangleq \{x \in \mathbb{R}^n : q_j(x) \ge 0, \quad j = 1, \dots, M\} \tag{5.15}$$

where q_j are real polynomials. Thus $\mathscr{V}_\beta$ and $\mathscr{U}$ are closed semialgebraic sets and the set containment constraint (5.14) is satisfied if there exists $\varepsilon > 0$, $p_i \ge 0$, $i = 1, \dots, N$, and $s_j \in \Sigma[x]$, $j = 1, \dots, M$, such that

$$\sum_{i=1}^{N} p_i V_i(x_i) - \beta - \varepsilon - \sum_{j=1}^{M} s_j(x) q_j(x) \in \Sigma[x]. \tag{5.16}$$

To see that (5.16) guarantees (5.14) note that $x \in \mathscr{U}$ implies $\sum_{j=1}^{M} s_j(x) q_j(x) \ge 0$ by definition of $\mathscr{U}$ and the fact that each s_j is SOS. Therefore, $V(x) - \beta - \varepsilon \ge 0$ which implies $V(x) \ge \beta + \varepsilon$, hence $x \in \overline{\mathscr{V}}_\beta$.

Proposition 5.2 *Suppose each subsystem G_i, defined in (2.1) and (2.2) with $f_i(0, 0) = 0$, $h_i(0, 0) = 0$, is dissipative with the quadratic supply rate (2.4) and storage function $V_i(\cdot)$. If there exist $\varepsilon > 0$, $p_i \ge 0$, $i = 1, \dots, N$, and $s_j \in \Sigma[x]$, $j = 1, \dots, M$, satisfying (5.16) and (5.6) with W as in (5.13), then trajectories starting from $x(0) = 0$ cannot intersect the unsafe set $\mathscr{U}$ for any $d(\cdot)$ with $\|d\|_2^2 \le \beta$.*

If the assumption $f_i(0, 0) = 0$, $h_i(0, 0) = 0$ is removed from Proposition 5.2 we must use the equilibrium independent properties of the subsystems. We assume an equilibrium x^* exists as in (3.1) with $M = M_{uy}$ and that each subsystem is EID with respect to quadratic supply rates given by X_i, $i = 1, \ldots, N$.

The safety constraint (5.16) must now be modified since the subsystem storage functions depend on the unknown equilibrium x^*. The unsafe set $\mathcal{U}$ may also depend on x^*; for example, we may consider the system safe if all trajectories remain within a distance of the equilibrium. We accommodate such scenarios with polynomials $q_j(x, x^*)$ that depend on x^* in (5.15).

The set containment constraint (5.14) is satisfied if there exists $\varepsilon > 0$, $p_i \geq 0$, $i = 1, \ldots, N$, $s_j \in \Sigma[x, \bar{x}]$, $j = 1, \ldots, M$, and $r_k \in \mathbb{R}[x_k, \bar{x}_k, u_k, \bar{u}_k]$, $k = 1, \ldots, N$ such that

$$\sum_{i=1}^{N} p_i V_i(x_i, \bar{x}_i) - \beta - \varepsilon - \sum_{j=1}^{M} s_j(x, \bar{x}) q_j(x, \bar{x})$$

$$- \sum_{k=1}^{N} r_k(x_k, \bar{x}_k, u_k, \bar{u}_k) f_k(\bar{x}_k, \bar{u}_k) \in \Sigma[x, \bar{x}, u, \bar{u}]. \tag{5.17}$$

Note that $\bar{x}$ and $\bar{u}$ in (5.17) are variables and not assumed to satisfy $f(\bar{x}, \bar{u}) = 0$. Instead, the r_k terms ensure that whenever $f(\bar{x}, \bar{u}) = 0$ then

$$\sum_{i=1}^{N} p_i V_i(x_i, \bar{x}_i) - \beta - \varepsilon - \sum_{j=1}^{M} s_j(x, \bar{x}) q_j(x, \bar{x}) \in \Sigma[x, \bar{x}]. \tag{5.18}$$

Therefore, we can remove the restriction $f_i(0, 0) = 0$, $h_i(0, 0) = 0$ from Proposition 5.2 by requiring that the subsystems be EID and the safety constraint (5.16) be replaced with (5.17). Then, trajectories starting from $x(0) = x^*$ cannot intersect the unsafe set $\mathcal{U}$ for any $d(\cdot)$ with $\|d\|_2^2 \leq \beta$.

It is straightforward to extend the results above to the case where the initial state belongs to a semialgebraic set rather than being located at the equilibrium. Suppose the initial state is contained in the set

$$\mathcal{I} \triangleq \{x \in \mathbb{R}^n : w_\ell(x) \geq 0, \quad \ell = 1, \ldots, L\} \tag{5.19}$$

where w_ℓ are real polynomials. If (5.6) holds and $\mathcal{I} \subset \mathcal{V}_\alpha$ then $x(t)$ is contained in the sublevel set $\mathcal{V}_{\alpha+\beta}$ for all $d(\cdot)$ with $\|d\|_2^2 \leq \beta$, $x(0) \in \mathcal{I}$, and $t \geq 0$. Using SOS techniques we can certify $\mathcal{I} \subset \mathcal{V}_\alpha$ if there exists $p_i \geq 0$, $i = 1, \ldots, N$, $t_\ell \in \Sigma[x, \bar{x}]$, $\ell = 1, \ldots, L$, and $r_k \in \mathbb{R}[x_k, \bar{x}_k, u_k, \bar{u}_k]$, $k = 1, \ldots, N$ satisfying

$$- \left(\sum_{i=1}^{N} p_i V_i(x_i) - \alpha \right) - \sum_{\ell=1}^{L} t_\ell(x, \bar{x}) w_\ell(x, \bar{x})$$

$$- \sum_{k=1}^{N} r_k(x_k, \bar{x}_k, u_k, \bar{u}_k) f_k(\bar{x}_k, \bar{u}_k) \in \Sigma[x, \bar{x}, u, \bar{u}]. \tag{5.20}$$

Therefore, the system is safe if the level set $\mathcal{V}_{\alpha+\beta}$ does not intersect the unsafe set $\mathcal{U}$. To guarantee this (5.17) must hold with β replaced by $\beta + \alpha$.

A similar safety certification procedure was developed in [1] for disturbances satisfying a bound on $d(t)$ for all t rather than in the L_2 norm sense. A direct application of sum-of-squares techniques to safety verification, without the compositional approach here, was reported in [2]. An overview of the broader literature on establishing invariant sets is given in [3].

5.3 Platoon Example Revisited

We illustrate the safety certification procedure above on the vehicle platoon example of Sect. 4.2. Recall that v_i, $i = 1, \ldots, N$, is the velocity of the i-th vehicle and z_l, $l = 1, \ldots, L$, is the relative position of the vehicles connected by the l-th link.

We consider an additive disturbance $d(t) \in \mathbb{R}^N$ on the velocity dynamics and wish to find a L_2 norm bound $\|d\|_2^2 \leq \beta$ such that the disturbance will not cause a collision. Thus we select the unsafe set to be

$$\mathcal{U} = \cup_{l=1,\ldots,L}\mathcal{U}_l \quad \text{where} \quad \mathcal{U}_l = \{(v, z) : |z_l| \leq \gamma\} \tag{5.21}$$

with a prescribed safety margin $\gamma > 0$, as depicted in Fig. 5.2.

Let the control law be as in (4.7) with $h_l(z_l) = (z_l - z_0)^{1/3}$, $l = 1, \ldots, L$, where $z_0 > 0$. Since h_l is increasing and onto, a unique equilibrium point exists as shown in Sect. 4.2.

Fig. 5.2 A cross section of the unsafe set (5.21) and a sublevel set $\mathcal{V}_\beta$ that certifies safety under disturbances satisfying $\|d\|_2^2 \leq \beta$

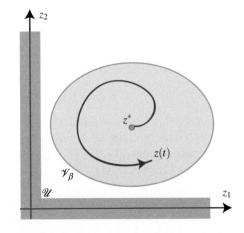

The subsystems mapping $u_i \mapsto v_i$ are given in (4.4). The storage functions

$$S_i(v_i, \bar{v}_i) = \frac{1}{2}(v_i - \bar{v}_i)^2, \quad i = 1, \ldots, N \tag{5.22}$$

certify that each subsystem is equilibrium independent output strictly passive since

$$\begin{aligned} \nabla_{v_i} S_i(v_i, \bar{v}_i) f_i(v_i, u_i) &= (v_i - \bar{v}_i)(-v_i + v_i^0 + u_i) \\ &= (v_i - \bar{v}_i)(-v_i + \bar{v}_i + u_i - \bar{u}_i) \\ &= \begin{bmatrix} u_i - \bar{u}_i \\ v_i - \bar{v}_i \end{bmatrix}^T \begin{bmatrix} 0 & 1/2 \\ 1/2 & -1 \end{bmatrix} \begin{bmatrix} u_i - \bar{u}_i \\ v_i - \bar{v}_i \end{bmatrix}, \end{aligned} \tag{5.23}$$

where we have used $f_i(\bar{v}_i, \bar{u}_i) = -\bar{v}_i + v_i^0 + \bar{u}_i = 0$ in the second equation.

The z_l subsystems are integrators with input w_l and output $h_l(z_l) = (z_l - z_0)^{1/3}$. The storage functions

$$R_l(z_l, \bar{z}_l) = \frac{3}{4}(z_l - z_0)^{4/3} - (z_l - z_0)(\bar{z}_l - z_0)^{1/3} + \frac{1}{4}(\bar{z}_l - z_0)^{4/3}, \quad l = 1, \ldots, L \tag{5.24}$$

certify equilibrium independent passivity since

$$\nabla_{z_l} R_l(z_l, \bar{z}_l) w_l = ((z_l - z_0)^{1/3} - (\bar{z}_l - z_0)^{1/3}) w_l \tag{5.25}$$

$$= \begin{bmatrix} w_l - \bar{w}_l \\ h_l(z_l) - h_l(\bar{z}_l) \end{bmatrix}^T \begin{bmatrix} 0 & 1/2 \\ 1/2 & 0 \end{bmatrix} \begin{bmatrix} w_l - \bar{w}_l \\ h_l(z_l) - h_l(\bar{z}_l) \end{bmatrix} \tag{5.26}$$

where $\bar{w}_l = 0$.

The composite storage function is

$$V(v, v^*, z, z^*) = \sum_{i=1}^{N} p_i S_i(v_i, v_i^*) + \sum_{l=1}^{L} p_{N+l} R_l(z_l, z_l^*) \tag{5.27}$$

and the weights p_i must satisfy (5.6) with W as in (5.13) to ensure dissipativity with the L_2 reachability supply rate. In addition, p_i must satisfy the set containment constraint (5.17). Since $\mathcal{U}$ is a union of the sets $\mathcal{U}_l$, it is necessary to include a constraint of the form (5.17) for each l.

To reduce the dimension of the problem we recall that the skew symmetric coupling of the subsystems suggests equal weights p_i. Indeed the choice $p_i = 4$ satisfies (5.6) with W as in (5.13). Thus we fix $p_i = 4$ and treat β as a decision variable in the set containment constraints.

As a numerical example consider a formation of $N = 3$ vehicles as in Fig. 4.2. The unsafe set $\mathcal{U} \triangleq \{z_1 : |z_1| \le 5\} \cup \{z_2 : |z_2| \le 5\}$ is the union of two sets; therefore, we include a constraint of the form (5.17) for each set. We let $v_1^0 = 9$, $v_2^0 = 10$, $v_3^0 = 11$, and $z_0 = 20$. Assuming the system is initialized at the equilibrium and a disturbance $d(\cdot)$ is applied to the third vehicle, we verified safety for all disturbances such that $\|d\|_2^2 \leq 52.0$.

Note that it is not obvious how to apply the SOS techniques to the functions h_l and R_l since they have fractional powers. To remedy this we replace $(z_l - z_0)^{1/3}$ and $(\bar{z}_l - z_0)^{1/3}$ with the auxillary variables y_l and $\bar{y}_l$, and include the polynomials equality constraints $y_l^3 = z_l - z_0$ and $\bar{y}_l^3 = \bar{z}_l - z_0$ in the SOS program. More information about applying SOS techniques to nonpolynomial systems can be found in [4].

References

1. Coogan, S., Arcak, M.: A dissipativity approach to safety verification for interconnected systems. IEEE Trans. Autom. Control **60**(6), 1722–1727 (2015). doi:10.1109/TAC.2014.2361595
2. Prajna, S., Jadbabaie, A., Pappas, G.: A framework for worst-case and stochastic safety verification using barrier certificates. IEEE Trans. Autom. Control **52**(8), 1415–1428 (2007)
3. Blanchini, F.: Set invariance in control. Automatica **35**(11), 1747–1767 (1999)
4. Papachristodoulou, A., Prajna, S.: Analysis of non-polynomial systems using the sum of squares decomposition. In: Henrion, D., Garulli, A. (eds.) Positive Polynomials in Control. Lecture Notes in Control and Information Science, vol. 312, pp. 23–43. Springer, Berlin (2005)

Chapter 6
Searching Over Subsystem Dissipativity Properties

6.1 Conical Combinations of Multiple Supply Rates

The stability and performance tests in earlier chapters used a fixed dissipativity property for each subsystem. This approach is effective when the interconnection structure suggests a compatible dissipativity property as in the case studies. However, in general, useful structural properties of the interconnection and relevant dissipativity properties may not be apparent.

A more flexible approach is to employ a combination of several dissipativity certificates known for each subsystem. Indeed, if a system is dissipative with respect to the supply rate and storage function pairs

$$(s_q(u, y), V_q(x)) \quad q = 1, \ldots, Q \tag{6.1}$$

then, by Definition 1.1, it is also dissipative with respect to any conical combination

$$\left(\sum_{q=1}^{Q} p_q s_q(u, y), \sum_{q=1}^{Q} p_q V_q(x) \right), \quad p_q \geq 0 \quad q = 1, \ldots, Q. \tag{6.2}$$

Thus, when each subsystem $i = 1, \ldots, N$ in the interconnection of Fig. 2.1 is dissipative with a set of quadratic supply rates given by

$$X_{i,q}, \quad q = 1, \ldots, Q_i,$$

we replace $\mathbf{X}(p_1 X_1, \ldots, p_N X_N)$ in the stability test (2.8) and performance test (5.6) with

$$\mathbf{X}\left(\sum_{q=1}^{Q_1} p_{1,q} X_{1,q}, \ldots, \sum_{q=1}^{Q_N} p_{N,q} X_{N,q} \right) \tag{6.3}$$

© The Author(s) 2016
M. Arcak et al., *Networks of Dissipative Systems*,
SpringerBriefs in Control, Automation and Robotics,
DOI 10.1007/978-3-319-29928-0_6

and leave the weights $p_{i,q}$ as decision variables.

As an illustration consider a negative feedback interconnection of two subsystems, with M as in (2.9). Suppose, we have a single dissipativity certificate for the first subsystem and two for the second subsystem:

$$X_1 = \begin{bmatrix} 1 & 1/2 \\ 1/2 & 0 \end{bmatrix} \quad X_{2,1} = \begin{bmatrix} -1 & 1/2 \\ 1/2 & 0 \end{bmatrix} \quad X_{2,2} = \begin{bmatrix} 1 & 0 \\ 0 & -1 \end{bmatrix}.$$

With $\mathbf{X}\left(p_1 X_1, \, p_{2,1} X_{2,1} + p_{2,2} X_{2,2}\right)$, the stability condition (2.8) becomes

$$\begin{bmatrix} p_{2,2} - p_{2,1} & (p_{2,1} - 1)/2 \\ (p_{2,1} - 1)/2 & 1 - p_{2,2} \end{bmatrix} \le 0 \tag{6.4}$$

where we have fixed $p_1 = 1$ since one of the decision variables can be factored out of (2.8). Note that (6.4) holds with the combination $p_{2,1} = p_{2,2} = 1$, but cannot hold when either $p_{2,1} = 0$ or $p_{2,2} = 0$. Thus, neither $X_{2,1}$ nor $X_{2,2}$ alone can prove the stability of the interconnection and a combination is essential.

6.2 Mediated Search for New Supply Rates

In this section, we take a more exhaustive approach and combine the stability and performance tests with a simultaneous search for feasible subsystem dissipativity properties. The supply rates $X_1, \ldots, X_N$ in the LMIs (2.8) and (5.6) are now decision variables instead of being fixed, and each X_i must satisfy the *local constraint*:

$$\nabla V_i(x_i)^T f_i(x_i, u_i) - \begin{bmatrix} u_i \\ h_i(x_i, u_i) \end{bmatrix}^T X_i \begin{bmatrix} u_i \\ h_i(x_i, u_i) \end{bmatrix} \le 0 \tag{6.5}$$

for all $x_i \in \mathbb{R}^{n_i}$, $u_i \in \mathbb{R}^{m_i}$, with an appropriate storage function $V_i(\cdot)$.

Since X_i is now a variable, and scaling both X_i and $V_i(\cdot)$ by $p_i \ge 0$ does not change (6.5), we drop the weights p_i from (2.8) and (5.6). We thus obtain the *global constraint* for performance:

$$\begin{bmatrix} M_{uy} & M_{ud} \\ I & 0 \\ 0 & I \\ M_{ey} & M_{ed} \end{bmatrix}^T \begin{bmatrix} \mathbf{X}(X_1, \ldots, X_N) & 0 \\ 0 & -W \end{bmatrix} \begin{bmatrix} M_{uy} & M_{ud} \\ I & 0 \\ 0 & I \\ M_{ey} & M_{ed} \end{bmatrix} \le 0 \tag{6.6}$$

where $\mathbf{X}(X_1, \ldots, X_N)$ is as defined in (2.7). The constraint for stability is the special case $W = 0$ and is not discussed separately.

Solving the combined feasibility problem (6.5)–(6.6) directly may be intractable for large networks, especially if the local problems (6.5) require sum-of-squares programming. Note, however, the subproblems (6.5) are coupled in (6.6) only by the supply rate variables X_i while the storage functions $V_i(\cdot)$ remain private. This sparse coupling allows us to decompose and solve (6.5)–(6.6) with scalable distributed optimization methods.

A particularly attractive method is the alternating direction method of multipliers (ADMM) which guarantees convergence under very mild assumptions [1]. For a general problem of the form:

$$
\begin{aligned}
\text{minimize} \quad & \phi(x) + \psi(z) \\
\text{subject to} \quad & Ax + Bz = c,
\end{aligned}
\tag{6.7}
$$

where x and z are vector decision variables, the ADMM updates are:

$$
x^{k+1} = \arg\min_x \ \phi(x) + \|Ax + Bz^k - c + s^k\|^2
\tag{6.8}
$$

$$
z^{k+1} = \arg\min_z \ \psi(z) + \|Ax^{k+1} + Bz - c + s^k\|^2
\tag{6.9}
$$

$$
s^{k+1} = s^k + Ax^{k+1} + Bz^{k+1} - c.
\tag{6.10}
$$

In particular, the variable s in (6.10) accumulates the deviation from the constraint $Ax + Bz = c$ as in integral control.

To bring the feasibility problem (6.5)–(6.6) to the canonical optimization form (6.7), we first define the indicator functions:

$$
\mathbb{I}_{\text{local},i}(X_i, V_i) := \begin{cases} 0 & \text{if } X_i, V_i \text{ satisfy (6.5)} \\ \infty & \text{otherwise} \end{cases}
\tag{6.11}
$$

$$
\mathbb{I}_{\text{global}}(X_1,\ldots, X_N) := \begin{cases} 0 & \text{if } X_1,\ldots, X_N \text{ satisfy (6.6)} \\ \infty & \text{otherwise.} \end{cases}
\tag{6.12}
$$

Next, we replace $X_1,\ldots, X_N$ in $\mathbb{I}_{\text{global}}$ with the auxiliary variables $Z_1,\ldots, Z_N$, and rewrite (6.5)–(6.6) as

$$
\begin{aligned}
\underset{X_i, V_i, Z_i, \ i=1,\ldots,N}{\text{minimize}} \quad & \sum_{i=1}^{N} \mathbb{I}_{\text{local},i}(X_i, V_i) + \mathbb{I}_{\text{global}}(Z_1,\ldots, Z_N) \\
\text{subject to} \quad & X_i - Z_i = 0 \quad \text{for} \quad i = 1,\ldots, N.
\end{aligned}
$$

The auxiliary variables $Z_1,\ldots, Z_N$ enabled the separation of the objective into $N+1$ independent functions. Thanks to this separation, the ADMM algorithm (6.8)–(6.10) takes the parallelized form below.

X-updates: For each i, solve the local problem

$$
X_i^{k+1} = \arg\min_{X \text{ s.t. (6.5) with } V \geq 0} \|X - Z_i^k + S_i^k\|_F^2
$$

where $\|\cdot\|_F$ represents the Frobenius norm.

Z-update: If $X_1^{k+1}, \ldots, X_N^{k+1}$ satisfy (6.6), then terminate. Otherwise, solve the global problem

$$Z_{1:N}^{k+1} = \arg\min_{(Z_1,\ldots,Z_N)\,\text{s.t. (6.6)}} \sum_{i=1}^{N} \left\| X_i^{k+1} - Z_i + S_i^k \right\|_F^2.$$

S-updates: Update S_i by

$$S_i^{k+1} = X_i^{k+1} - Z_i^{k+1} + S_i^k$$

and return to the X-updates.

For each subsystem, this algorithm solves an optimization problem certifying dissipativity with a supply rate X_i close to the Z_i proposed by the global problem. The global problem first checks if the constraint (6.6) is satisfied with the updated supply rates X_i. If not, it solves an optimization problem to propose new supply rates Z_i, close to X_i, that satisfy (6.6). Thus, the global problem mediates between the local searches for supply rates to find a feasible combination.

For equilibrium independent certification of stability and performance, the global constraint (6.6) remains unchanged if the subsystem dissipativity assumption is replaced with its equilibrium independent form. Thus, the only change needed in the ADMM algorithm above is to adapt the X-updates to local EID constraints.

Other distributed optimization methods are applicable to this formulation. Subgradient methods combined with dual decomposition [2] were employed for stability certification from L_2 gain properties of the subsystems [3], and later extended to general dissipativity [4]. Unlike ADMM, this method calls for careful tuning of the stepsize schedule and regularization parameter. Projection methods [5, 6] are also applicable; however, the convergence rates may be very slow [4].

A Relaxed Exit Criterion

Before the Z-update the algorithm checks if $X_1^{k+1}, \ldots, X_N^{k+1}$ satisfy the global constraint (6.6). If so, performance is certified and the algorithm terminates.

Since the ADMM algorithm generates a sequence of supply rates X_i^q, $q = 1, \ldots, k+1$ whose conical combinations are also valid supply rates (Sect. 6.1), we can instead check if (6.6) is satisfied with

$$\mathbf{X}\left(\sum_{q=1}^{k+1} p_{1,q} X_{1,q}, \ldots, \sum_{q=1}^{k+1} p_{N,q} X_{N,q} \right) \tag{6.13}$$

where the weights $p_{i,q} \geq 0$ are decision variables. Alternatively, one may consider a subset of recent supply rates rather than the whole sequence $q = 1, \ldots, k+1$.

This modification does not affect the iterations of the ADMM algorithm, only the exit criterion. Thus, the algorithm is still guaranteed to converge, but the number of iterations can be greatly reduced. As an example, an interconnection of 100 two-state nonlinear single-input single-output systems was generated. For each test the subsystem parameters and interconnection were chosen randomly but constrained so that the system had L_2 gain less then or equal to one. On 50 instances of this problem the standard ADMM algorithm required on average 14.7 iterations. With the modified exit criterion this average dropped to 4.8.

References

1. Boyd, S., Parikh, N., Chu, E., Peleato, B., Eckstein, J.: Distributed optimization and statistical learning via the alternating direction method of multipliers. Found. Trends Mach. Learn. **3**(1), 1–122 (2011)
2. Nedić, A., Ozdaglar, A.: Cooperative distributed multi-agent optimization. In: Palomar, D.P., Eldar, Y.C. (eds.) Convex Optimization in Signal Processing and Communications, pp. 340–386. Cambridge University Press (2009)
3. Topcu, U., Packard, A., Murray, R.: Compositional stability analysis based on dual decomposition. In: IEEE Conference on Decision and Control, pp. 1175–1180 (2009)
4. Meissen, C., Lessard, L., Packard, A.: Performance certification of interconnected systems using decomposition techniques. In: American Control Conference, pp. 5030–5036 (2014)
5. Gubin, L.G., Polyak, B.T., Raik, E.V.: The method of projections for finding the common point of convex sets. USSR Comput. Math. Math. Phys. **7**(6), 1–24 (1967)
6. Bauschke, H., Borwein, J.: Dykstra's alternating projection algorithm for two sets. J. Approx. Theory **79**(3), 418–443 (1994)

Chapter 7
Symmetry Reduction

7.1 Reduction for Stability Certification

We revisit the stability certification problem and exploit the symmetries in the inter-connection of Fig. 2.1 to reduce the number of decision variables. To avoid cumbersome notation we assume single-input single-output subsystems, i.e., $M \in \mathbb{R}^{N \times N}$.

To characterize symmetries of M we define a permutation matrix R satisfying

$$RM = MR \tag{7.1}$$

to be an *automorphism* of M. If we permute the indices of the subsystems according to such R, the interconnection does not change (it *morphes* into *itself*) because (7.1) ensures that the inputs $\tilde{u} = Ru$ and outputs $\tilde{y} = Ry$, relabeled with the new indices, still satisfy $\tilde{u} = M\tilde{y}$.

As an illustration, consider the cyclic interconnection (2.13) with $N = 6, \delta_i = -1$ when i is odd, and $\delta_i = +1$ when i is even; see the incidence graph in Fig. 7.1 (left). A permutation that rotates the indices by two nodes is an automorphism because the interconnection remains unchanged (right). By contrast, rotating the indices by one node would change the signs of the edges connecting any two nodes.

The set of all automorphisms of M forms a group, denoted

$$\text{Aut}(M) = \{R \text{ such that } (7.1) \text{ holds}\}. \tag{7.2}$$

Given this automorphism group we define the *orbit* of node $i \in \{1, \ldots, N\}$ to be the set of all nodes j such that some element R permutes i to j. That is,

$$O_i = \{j \in \{1, \ldots, N\} \mid Rq_i = q_j \quad \text{for some } R \in \text{Aut}(M)\} \tag{7.3}$$

© The Author(s) 2016
M. Arcak et al., *Networks of Dissipative Systems*,
SpringerBriefs in Control, Automation and Robotics,
DOI 10.1007/978-3-319-29928-0_7

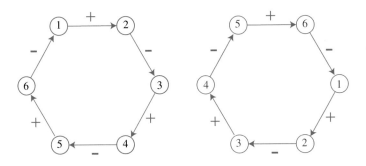

Fig. 7.1 For the interconnection depicted on the *left*, a permutation that rotates the indices by two nodes (*right*) is an automorphism because the edges connecting the nodes are unchanged

where $q_i = \mathbb{R}^N$ is the ith unit vector. The orbits partition the nodes $1, \ldots, N$ into equivalence classes, defined by the relation

$$i \sim j \quad \text{if } j \in O_i, \tag{7.4}$$

where nodes in the same class can be reached from one another by an automorphism. The two distinct orbits in Fig. 7.1 are $\{1, 3, 5\}$ and $\{2, 4, 6\}$.

The following theorem states that, if the subsystems (nodes) on the same orbit have identical supply rates, $X_i = X_j$ when $i \sim j$, then taking identical weights $p_i = p_j$ for $i \sim j$ does not change the feasibility of (2.8). Thus, we need one decision variable per orbit rather than one for each node.

Theorem 7.1 *Given $X_1, \ldots, X_N$ such that $X_i = X_j$ when $i \sim j$, if (2.8) holds with weights p_i, $i = 1, \ldots, N$, then it also holds with*

$$\bar{p}_i = \frac{1}{|O_i|} \sum_{j \in O_i} p_j \quad i = 1, \ldots, N \tag{7.5}$$

where $|O_i|$ is the number of elements in (7.3). In particular, $\bar{p}_i = \bar{p}_j$ for $i \sim j$.

Proof We will prove the implication

$$\begin{bmatrix} M \\ I \end{bmatrix}^T \mathbf{X}(Y_1, \ldots, Y_N) \begin{bmatrix} M \\ I \end{bmatrix} \leq 0 \quad \Rightarrow \quad \begin{bmatrix} M \\ I \end{bmatrix}^T \mathbf{X}(\bar{Y}_1, \ldots, \bar{Y}_N) \begin{bmatrix} M \\ I \end{bmatrix} \leq 0 \tag{7.6}$$

where

$$\bar{Y}_i = \frac{1}{|O_i|} \sum_{j \in O_i} Y_j. \tag{7.7}$$

The theorem follows from this implication by setting $Y_i = p_i X_i$. In particular, the assumption that $X_j = X_i$ for all $j \in O_i$ reduces (7.7) to $\bar{p}_i X_i$.

Let $R \in \mathrm{Aut}(M)$ and note that the left-hand side of (7.6) implies

$$R^T \begin{bmatrix} M \\ I \end{bmatrix}^T \mathbf{X}(Y_1, \dots, Y_N) \begin{bmatrix} M \\ I \end{bmatrix} R \leq 0 \tag{7.8}$$

which, by (7.1), is identical to

$$\begin{bmatrix} M \\ I \end{bmatrix}^T \begin{bmatrix} R & 0 \\ 0 & R \end{bmatrix}^T \mathbf{X}(Y_1, \dots, Y_N) \begin{bmatrix} R & 0 \\ 0 & R \end{bmatrix} \begin{bmatrix} M \\ I \end{bmatrix} \leq 0. \tag{7.9}$$

It follows from the definition of $\mathbf{X}(Y_1, \dots, Y_N)$ in (2.7) that

$$\begin{bmatrix} R & 0 \\ 0 & R \end{bmatrix}^T \mathbf{X}(Y_1, \dots, Y_N) \begin{bmatrix} R & 0 \\ 0 & R \end{bmatrix} = \mathbf{X}(Y_{R(1)}, \dots, Y_{R(N)}) \tag{7.10}$$

where $R(i)$ denotes the index to which i gets permuted by the automorphism R. Thus,

$$\begin{bmatrix} M \\ I \end{bmatrix}^T \mathbf{X}(Y_{R(1)}, \dots, Y_{R(N)}) \begin{bmatrix} M \\ I \end{bmatrix} \leq 0. \tag{7.11}$$

Averaging the expression on the left over $\mathrm{Aut}(M)$ (that is, adding over $R \in \mathrm{Aut}(M)$ and dividing by $|\mathrm{Aut}(M)|$) we obtain the right-hand side of (7.6). □

The theorem above holds for any subset of automorphisms that forms a group. This generality is important for applications where the full automorphism group is difficult to compute but a subset representing a particular symmetry is apparent. However, in this case the reduction may not be as extensive.

Enriching Symmetries for Further Reduction
The proposition below shows that transformations of the form

$$\hat{M} = D^{-1} M D \tag{7.12}$$

where $D \in \mathbb{C}^{N \times N}$ is diagonal do not change the feasibility of (2.8). We apply such transformations to enrich the symmetries in M thereby reducing the number of orbits and the corresponding decision variables in (2.8).

As an example, for the cyclic interconnection in Fig. 7.1 the choice of D specified in the next section yields identical edge weights $(=e^{j\pi/6})$ which means that all rotations are now automorphisms and the number of orbits is reduced to one.

Proposition 7.1 *Let $\hat{M}$ be as in (7.12) where D is a diagonal matrix with entries $d_i \in \mathbb{C}$, $d_i \neq 0$, $i = 1, \ldots, N$. Then the LMI (2.8) is equivalent to*

$$\begin{bmatrix} \hat{M} \\ I \end{bmatrix}^* \mathbf{X}(\hat{p}_1 X_1, \ldots, \hat{p}_N X_N) \begin{bmatrix} \hat{M} \\ I \end{bmatrix} \leq 0 \qquad (7.13)$$

where $\hat{p}_i = |d_i|^2 p_i$. Thus, if there exists $p_i > 0$ satisfying (2.8) then there exist $\hat{p}_i > 0$ satisfying (7.13) and vice versa.

Proof Multiplying (2.8) from the left by D^* and from the right by D we get

$$D^* \begin{bmatrix} M \\ I \end{bmatrix}^* \mathbf{X}(p_1 X_1, \ldots, p_N X_N) \begin{bmatrix} M \\ I \end{bmatrix} D \leq 0 \qquad (7.14)$$

which, by (7.12), identical to

$$\begin{bmatrix} \hat{M} \\ I \end{bmatrix}^* \underbrace{\begin{bmatrix} D^* & 0 \\ 0 & D^* \end{bmatrix} \mathbf{X}(p_1 X_1, \ldots, p_N X_N) \begin{bmatrix} D & 0 \\ 0 & D \end{bmatrix}}_{=\mathbf{X}(|d_1|^2 p_1 X_1, \ldots, |d_N|^2 p_N X_N)} \begin{bmatrix} \hat{M} \\ I \end{bmatrix} \leq 0. \qquad (7.15)$$

$\square$

7.2 Cyclic Interconnections Revisited

We consider again the cyclic interconnection

$$M = \begin{bmatrix} 0 & \cdots & 0 & \delta_1 \\ \delta_2 & 0 & \cdots & 0 \\ \vdots & \ddots & \ddots & \vdots \\ 0 & \cdots & \delta_N & 0 \end{bmatrix} \qquad (7.16)$$

of output strictly passive systems with supply rate $s_i(u_i, y_i) = u_i y_i - \varepsilon_i y_i^2$, $\varepsilon_i > 0$. To examine the feasibility of the stability criterion (2.8) we first define

$$\tilde{u}_i \triangleq \varepsilon_i^{-1} u_i, \quad \tilde{s}_i(\tilde{u}_i, y_i) \triangleq \varepsilon_i^{-1} s_i(u_i, y_i) = \tilde{u}_i y_i - y_i^2,$$

so that each subsystem has identical supply rate given by

$$\tilde{X}_i = \begin{bmatrix} 0 & 1/2 \\ 1/2 & -1 \end{bmatrix} \qquad (7.17)$$

and the parameters ε_i are absorbed into the interconnection equation $\tilde{u} = \tilde{M} y$ where $\tilde{M}$ is specified in (7.19) below.

Next, we note that a transformation of the form (7.12) with diagonal entries

$$d_1 = 1, \quad d_i = d_{i-1} \frac{\delta_i}{\varepsilon_i} \frac{1}{r} \quad i = 2, \ldots, N, \quad r \triangleq \left(\frac{\delta_1 \ldots \delta_N}{\varepsilon_1 \ldots \varepsilon_N} \right)^{1/N} \tag{7.18}$$

endows the interconnection with rotational symmetry:

$$\tilde{M} = \begin{bmatrix} 0 & \cdots & 0 & \frac{\delta_1}{\varepsilon_1} \\ \frac{\delta_2}{\varepsilon_2} & 0 & \cdots & 0 \\ \vdots & \ddots & \ddots & \vdots \\ 0 & \cdots & \frac{\delta_N}{\varepsilon_N} & 0 \end{bmatrix} \quad \hat{M} = D^{-1} \tilde{M} D = \begin{bmatrix} 0 & \cdots & 0 & r \\ r & 0 & \cdots & 0 \\ \vdots & \ddots & \ddots & \vdots \\ 0 & \cdots & r & 0 \end{bmatrix}. \tag{7.19}$$

Thus the entire set $\{1, \ldots, N\}$ is a single orbit under the automorphism group of $\hat{M}$.

By Proposition 7.1 the feasibility of the LMI (2.8) is equivalent to that of (7.13), and by Theorem 7.1 taking equal weights $\hat{p}_1 = \cdots = \hat{p}_N$, say $= 1$, does not restrict feasibility. Substituting

$$\mathbf{X}(\tilde{X}_1, \ldots, \tilde{X}_N) = \begin{bmatrix} 0 & \frac{1}{2} I \\ \frac{1}{2} I & -I \end{bmatrix} \tag{7.20}$$

in (7.13) we get the following necessary and sufficient feasibility condition for (2.8):

$$\frac{1}{2} \hat{M} + \frac{1}{2} \hat{M}^* - I \leq 0. \tag{7.21}$$

Note that (7.21) defines a *circulant matrix* whose first row is

$$\begin{bmatrix} -1 & \frac{1}{2} r^* & 0 & \cdots & 0 & \frac{1}{2} r \end{bmatrix} \tag{7.22}$$

and the subsequent rows are obtained by shifting the entries to the right with a wrap around from the Nth entry to the first. The eigenvalues of circulant matrices are the discrete Fourier transform coefficients of the first row [1] which, for (7.22), are

$$\lambda_k = -1 + \frac{1}{2} r^* e^{-j \frac{2\pi}{N} k} + \frac{1}{2} r e^{j \frac{2\pi}{N} k} \quad k = 1, \ldots, N. \tag{7.23}$$

Following the definition of r in (7.18), we substitute $r = |r| e^{j\pi/N}$ when $\delta_1 \ldots \delta_N < 0$, and $r = |r|$ when $\delta_1 \ldots \delta_N \geq 0$, obtaining

$$\lambda_k = \begin{cases} -1 + |r| \cos\left(\frac{\pi}{N} + \frac{2\pi}{N} k \right) & \text{when } \delta_1 \ldots \delta_N < 0 \\ -1 + |r| \cos\left(\frac{2\pi}{N} k \right) & \text{when } \delta_1 \ldots \delta_N \geq 0. \end{cases} \tag{7.24}$$

Since $\lambda_k \leq \lambda_N, k = 1, \ldots, N-1$, (7.21) is equivalent to $\lambda_N \leq 0$, that is

$$|r| \leq \begin{cases} \sec(\pi/N) & \text{when } \delta_1 \ldots \delta_N < 0 \\ 1 & \text{when } \delta_1 \ldots \delta_N \geq 0. \end{cases} \tag{7.25}$$

We summarize the result in the following proposition which recovers (2.14) when $\delta_1 \ldots \delta_N = -1$ as in (2.13).

Proposition 7.2 *Consider systems with supply rates $s_i(u_i, y_i) = u_i y_i - \varepsilon_i y_i^2$, $\varepsilon_i > 0, i = 1, \ldots, N$, interconnected according to (7.16). There exists $p_i > 0$, $i = 1, \ldots, N$, satisfying the stability criterion (2.8) if and only if*

$$|r|^N = \frac{|\delta_1 \ldots \delta_N|}{\varepsilon_1 \ldots \varepsilon_N} \leq \begin{cases} \sec^N(\pi/N) \text{ when } \delta_1 \ldots \delta_N < 0 \\ 1 \qquad\qquad \text{when } \delta_1 \ldots \delta_N \geq 0. \end{cases} \tag{7.26}$$

7.3 Reduction for Performance Certification

We now consider the interconnection in Fig. 5.1 with disturbance $d \in \mathbb{R}^m$, performance output $e \in \mathbb{R}^p$, and input and output vectors $u \in \mathbb{R}^N$, $y \in \mathbb{R}^N$ for the concatenation of N single-input single-output systems. The interconnection matrix is

$$\overline{M} = \begin{bmatrix} M_{uy} & M_{ud} \\ M_{ey} & M_{ed} \end{bmatrix} \tag{7.27}$$

with blocks $M_{uy} \in \mathbb{R}^{N \times N}$, $M_{ud} \in \mathbb{R}^{N \times m}$, $M_{ey} \in \mathbb{R}^{p \times N}$, $M_{ed} \in \mathbb{R}^{p \times m}$.

We generalize the notion of automorphism in Sect. 7.1 as follows:

Definition 7.1 The triplet (R, R_d, R_e) of permutation matrices $R \in \mathbb{R}^{N \times N}$, $R_d \in \mathbb{R}^{m \times m}$, $R_e \in \mathbb{R}^{p \times p}$ is an *automorphism* of $\overline{M}$ if

$$\overline{M} \begin{bmatrix} R & 0 \\ 0 & R_d \end{bmatrix} = \begin{bmatrix} R & 0 \\ 0 & R_e \end{bmatrix} \overline{M}. \tag{7.28}$$

This definition encompasses the one in Sect. 7.1 because (7.28) implies $R M_{uy} = M_{uy} R$, where M_{uy} plays the role of M in (7.1). However, we now ask that the permutation R be matched with a simultaneous permutation R_d of disturbances and

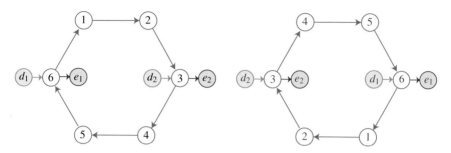

Fig. 7.2 An automorphism (R, R_d, R_e) where R rotates the nodes $1, \ldots, 6$ by three, R_d permutes d_1 with d_2, and R_e permutes e_1 with e_2. The interconnection is unchanged as shown on the *right*

R_e of performance variables that together leave the interconnection invariant. An example is shown in Fig. 7.2 (left) where M_{uy} has the form of $\hat{M}$ in (7.19),

$$M_{ey} = M_{ud}^T = \begin{bmatrix} 0\,0\,0\,0\,0\,1 \\ 0\,0\,1\,0\,0\,0 \end{bmatrix} \quad \text{and} \quad M_{ed} = 0. \tag{7.29}$$

All permutations R that rotate the nodes $1, \ldots, 6$ satisfy $R M_{uy} = M_{uy} R$. However, only rotation by three nodes, matched with a simultaneous permutation of d_1 with d_2 and e_1 with e_2, leaves the interconnection unchanged (right).

The set of all automorphisms defines the *automorphism group* $\mathrm{Aut}(\overline{M})$ and the *orbit* of node $i \in \{1, \ldots, N\}$ under this group is

$$O_i = \{j \in \{1, \ldots, N\} \mid R q_i = q_j \quad \text{for some } (R, R_d, R_e) \in \mathrm{Aut}(\overline{M})\}. \tag{7.30}$$

As before, the orbits partition $\{1, \ldots, N\}$ into equivalence classes with the relation $i \sim j$ indicating $j \in O_i$. The orbits in Fig. 7.2 are $\{1, 4\}$, $\{2, 5\}$, and $\{3, 6\}$.

We propose a reduction of the decision variables in the performance test (5.6) that mimics the reduction suggested in Theorem 7.1 for the stability test (2.8). For this extension we stipulate that the performance supply rate

$$\begin{bmatrix} d \\ e \end{bmatrix}^T W \begin{bmatrix} d \\ e \end{bmatrix} \tag{7.31}$$

be invariant under $\mathrm{Aut}(\overline{M})$, that is

$$W \begin{bmatrix} R_d & 0 \\ 0 & R_e \end{bmatrix} = \begin{bmatrix} R_d & 0 \\ 0 & R_e \end{bmatrix} W \quad \text{for all } (R, R_d, R_e) \in \mathrm{Aut}(\overline{M}). \tag{7.32}$$

For the example of Fig. 7.2, the L_2 gain supply rate $\gamma_1^2 d_1^2 + \gamma_2^2 d_2^2 - e_1^2 - e_2^2$ satisfies this condition if $\gamma_1 = \gamma_2$.

If the performance criterion satisfies this condition and the subsystems on the same orbit have identical supply rates, then taking identical weights $p_i = p_j$ for $i \sim j$ does not change the feasibility of the performance test (5.6). Thus, we can apply this test with one decision variable per orbit.

> **Theorem 7.2** *Suppose $X_1, \ldots, X_N$ satisfy $X_i = X_j$ when $i \sim j$ and W satisfies (7.32). If (5.6) holds with weights p_i, $i = 1, \ldots, N$, then it also holds with*
>
> $$\bar{p}_i = \frac{1}{|O_i|} \sum_{j \in O_i} p_j \quad i = 1, \ldots, N \qquad (7.33)$$
>
> *where $|O_i|$ is the number of elements in (7.30). In particular, $\bar{p}_i = \bar{p}_j$ for $i \sim j$.*

The proof is provided in [2] and follows closely the proof of Theorem 7.1 above. Similarly, an extension of Proposition 7.1 guarantees that the feasibility of the performance test (5.6) is unchanged under the transformation

$$\begin{bmatrix} \hat{M}_{uy} & \hat{M}_{ud} \\ \hat{M}_{ey} & \hat{M}_{ed} \end{bmatrix} = \begin{bmatrix} D^{-1} & 0 \\ 0 & D_e^{-1} \end{bmatrix} \begin{bmatrix} M_{uy} & M_{ud} \\ M_{ey} & M_{ed} \end{bmatrix} \begin{bmatrix} D & 0 \\ 0 & D_d \end{bmatrix} \quad \hat{W} = \begin{bmatrix} D_d & 0 \\ 0 & D_e \end{bmatrix}^* W \begin{bmatrix} D_d & 0 \\ 0 & D_e \end{bmatrix}$$

where $D \in \mathbb{C}^{N \times N}$, $D_e \in \mathbb{C}^{p \times p}$, $D_d \in \mathbb{C}^{m \times m}$ are diagonal and invertible. Such transformations are useful for generating symmetries that can then be used for a reduction in the number of decision variables. The computational benefits of the symmetry reduction above are studied in detail in [2].

Finally, we note that incorporating the symmetry reduction in the ADMM algorithm in Sect. 6.2 is possible with minor modifications. In this case, we do not assume that subsystems on the same orbit have identical supply rates, but rather enforce this condition. The minimization in the Z update is performed subject to the constraint $Z_i = Z_j$ for $i \sim j$; the X and S updates remain the same. The algorithm is terminated after the Z update if $Z_1, \ldots, Z_N$ satisfy the local constraints (6.2).

References

1. Davis, P.: Circulant Matrices. Wiley (1979)
2. Rufino Ferreira, A., Meissen, C., Arcak, M., Packard, A.: Symmetry Reduction for Performance Certification of Interconnected Systems (submitted)

Chapter 8
Dissipativity with Dynamic Supply Rates

8.1 Generalizing the Notion of Dissipativity

We now define a generalized notion of dissipativity that incorporates more information about a dynamical system than the standard form in Chap. 1. For this generalization we augment the model (1.1) and (1.2) with a stable linear system

$$\frac{d}{dt}\eta(t) = A\eta(t) + B\begin{bmatrix} u(t) \\ y(t) \end{bmatrix} \quad \eta(t) \in \mathbb{R}^{n'} \tag{8.1}$$

$$z(t) = C\eta(t) + D\begin{bmatrix} u(t) \\ y(t) \end{bmatrix} \quad z(t) \in \mathbb{R}^{p'} \tag{8.2}$$

that serves as a virtual filter for the inputs and outputs. The dimensions of η and z as
well as the choice of A, B, C, D depend on the dynamical properties of the system
(1.1) and (1.2) one would like to capture (Fig. 8.1).

> **Definition 8.1** The system (1.1) and (1.2) is dissipative with respect to the
> **dynamic supply rate** $z^T X z$ where z is the output of the auxiliary system (8.1)
> and (8.2) and X is a real symmetric matrix if there exists a storage function
> $V : \mathbb{R}^n \times \mathbb{R}^{n'} \mapsto \mathbb{R}$ such that $V(0, 0) = 0$, $V(x, \eta) \geq 0 \ \forall x, \eta$, and
>
> $$V(x(\tau), \eta(\tau)) - V(x(0), \eta(0)) \leq \int_0^\tau z(t)^T X z(t) dt \tag{8.3}$$
>
> for every input signal $u(\cdot)$ and every $\tau \geq 0$ in the interval of existence of the
> solution $x(t)$.

© The Author(s) 2016
M. Arcak et al., *Networks of Dissipative Systems*,
SpringerBriefs in Control, Automation and Robotics,
DOI 10.1007/978-3-319-29928-0_8

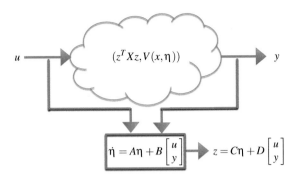

Fig. 8.1 Definition 8.1 generalizes the notion of dissipativity in Chap. 1 to allow for a dynamic supply rate $z^T X z$ where z is a filtered version of the vector of inputs and outputs. This generalization incorporates more detailed information from the underlying dynamical system. The earlier definition is the special case where $C = 0$, $D = I$

The standard form of dissipativity with a quadratic supply rate is a special case with $D = I$ and $C = 0$, that is,

$$z = \begin{bmatrix} u \\ y \end{bmatrix}.$$

For a continuously differentiable storage function $V(\cdot, \cdot)$, (8.3) is equivalent to

$$\nabla_x V(x, \eta)^T f(x, u) + \nabla_\eta V(x, \eta)^T \left(A\eta + B \begin{bmatrix} u \\ h(x, u) \end{bmatrix} \right)$$
$$\leq \left(C\eta + D \begin{bmatrix} u \\ h(x, u) \end{bmatrix} \right)^T X \left(C\eta + D \begin{bmatrix} u \\ h(x, u) \end{bmatrix} \right) \forall x \in \mathbb{R}^n, \ \eta \in \mathbb{R}^{n'}, \ u \in \mathbb{R}^m.$$
$$(8.4)$$

Example 8.1 The scalar system

$$\frac{dx(t)}{dt} = -\alpha x(t) + u(t) \quad \alpha > 0, \qquad y(t) = \gamma x(t) \quad \gamma > 0, \qquad (8.5)$$

is dissipative with supply rate $z^T \begin{bmatrix} 0 & 1/2 \\ 1/2 & -\varepsilon \end{bmatrix} z$ for some $\varepsilon > 0$ when z is generated by

$$\frac{d\eta(t)}{dt} = -\eta(t) + u(t) \qquad (8.6)$$

$$z(t) = \begin{bmatrix} -\beta\eta(t) + u(t) \\ y(t) \end{bmatrix} \quad \beta < \min\{\alpha, 1\}. \qquad (8.7)$$

The proof follows by showing output strict passivity of the (x, η) system with input $\hat{u} = -\beta\eta + u$ and output $y = \gamma x$. When $\alpha \neq 1$ the new variables $\chi_1 = \frac{\gamma}{1-\alpha}(x - \eta)$, $\chi_2 = \frac{\gamma}{1-\alpha}(-\alpha x + \eta)$ satisfy

$$\frac{d}{dt}\begin{bmatrix} \chi_1(t) \\ \chi_2(t) \end{bmatrix} = \begin{bmatrix} 0 & 1 \\ -\alpha(1-\beta) & -(1+\alpha-\beta) \end{bmatrix}\begin{bmatrix} \chi_1(t) \\ \chi_2(t) \end{bmatrix} + \begin{bmatrix} 0 \\ \gamma \end{bmatrix}\hat{u}(t) \qquad (8.8)$$

$$y(t) = \begin{bmatrix} 1 & 1 \end{bmatrix}\begin{bmatrix} \chi_1(t) \\ \chi_2(t) \end{bmatrix} \qquad (8.9)$$

which is of the form in Example 1.3 with $\ell = \alpha(1-\beta)$, $k = 1+\alpha-\beta$, and $\mu = 1$. Since $\beta < \min\{\alpha, 1\}$ we have $\ell > 0$ and $k > \mu > 0$; thus, from Example 1.3, the augmented (x, η) system is output strictly passive. When $\alpha = 1$, the augmented system cannot be brought to the form of Example 1.3 but can again be shown to be output strictly passive by showing the existence of a $P > 0$ satisfying (1.25).

Note that the choice $\beta = 0$ in (8.7) implies the output strict passivity of (8.5); the full class of filters with $\beta < \min\{\alpha, 1\}$ provides a more detailed description of the input/output behavior of (8.5).

Example 8.2 The previous example derived a class of filters that preserve an existing passivity property. In this example we characterize filters that attain passivity when combined with a system that lacks this property.

Consider the model

$$\frac{dx_1(t)}{dt} = x_2(t)$$

$$\frac{dx_2(t)}{dt} = -x_1(t) - kx_2(t) + u(t) \quad k \in (0, 1) \qquad (8.10)$$

$$y(t) = x_1(t) + x_2(t)$$

which violates the necessary condition for passivity in Example 1.3 because $k < 1$. We introduce the filter

$$\frac{d\eta(t)}{dt} = -\eta(t) + y(t) \qquad (8.11)$$

$$\hat{y}(t) = -\beta\eta(t) + y(t) \qquad (8.12)$$

and combine with the system equations above using the new variable $\chi_3 \triangleq \eta - x_1$:

$$\frac{dx_1(t)}{dt} = x_2(t)$$

$$\frac{dx_2(t)}{dt} = -x_1(t) - kx_2(t) + u(t) \quad k \in (0, 1) \qquad (8.13)$$

$$\frac{d\chi_3(t)}{dt} = -\chi_3(t)$$

$$\hat{y}(t) = (1-\beta)x_1(t) + x_2(t) - \beta\chi_3(t).$$

We then refer to Example 1.4 and examine

$$A = \begin{bmatrix} 0 & 1 \\ -1 & -k \end{bmatrix} \quad B = \begin{bmatrix} 0 \\ 1 \end{bmatrix} \quad C = \begin{bmatrix} (1-\beta) & 1 \end{bmatrix}$$

which excludes the uncontrollable χ_3 subsystem.

If we choose $\beta > 1-k$, it follows from Example 1.3 that there exists $P = P^T > 0$ satisfying (1.25). Then Example 1.4 implies that there exists $\hat{P} = \hat{P}^T > 0$ satisfying (1.27), thus certifying passivity of the augmented system (8.13). We conclude that system (8.10) is dissipative with supply rate

$$z^T \begin{bmatrix} 0 & 1/2 \\ 1/2 & 0 \end{bmatrix} z \quad \text{where} \quad z = \begin{bmatrix} u \\ \hat{y} \end{bmatrix} = \begin{bmatrix} u \\ \beta\eta + y \end{bmatrix}, \quad \beta > 1-k.$$

8.2   Stability of Interconnections

We revisit the interconnection in Fig. 2.1 and augment the subsystem models (2.1) and (2.2), $f_i(0,0) = 0$, $h_i(0,0) = 0$, with stable linear systems

$$\frac{d}{dt}\eta_i(t) = A_i\eta(t) + B_i \begin{bmatrix} u_i(t) \\ y_i(t) \end{bmatrix} \quad \eta_i(t) \in \mathbb{R}^{n'_i} \tag{8.14}$$

$$z_i(t) = C_i\eta_i(t) + D_i \begin{bmatrix} u_i(t) \\ y_i(t) \end{bmatrix} \quad z_i(t) \in \mathbb{R}^{p'_i}. \tag{8.15}$$

We then assume each subsystem is dissipative with a positive definite, continuously differentiable storage function $V_i(\cdot, \cdot)$ and supply rate $z_i^T X_i z_i$, that is

$$\nabla_{x_i} V_i(x_i, \eta_i)^T f_i(x_i, u_i) + \nabla_{\eta_i} V_i(x_i, \eta_i)^T \left(A_i\eta + B_i \begin{bmatrix} u_i \\ y_i \end{bmatrix} \right) \le z_i^T X_i z_i. \tag{8.16}$$

Defining A, B, C, D to be block diagonal matrices comprised of A_i, B_i, C_i, D_i, $i = 1, \ldots, N$, we lump (8.14) and (8.15) into a single auxiliary system

$$\frac{d}{dt}\eta(t) = A\eta(t) + BS \begin{bmatrix} u(t) \\ y(t) \end{bmatrix} = A\eta(t) + BS \begin{bmatrix} M \\ I \end{bmatrix} y(t) \tag{8.17}$$

$$z(t) = C\eta(t) + DS \begin{bmatrix} u(t) \\ y(t) \end{bmatrix} = C\eta(t) + DS \begin{bmatrix} M \\ I \end{bmatrix} y(t) \tag{8.18}$$

where M is the interconnection matrix and S is a permutation matrix such that

$$
S \begin{bmatrix} u_1 \\ \vdots \\ u_N \\ y_1 \\ \vdots \\ y_N \end{bmatrix} = \begin{bmatrix} u_1 \\ y_1 \\ \vdots \\ u_N \\ y_N \end{bmatrix}. \tag{8.19}
$$

Next we search for a Lyapunov function of the form

$$
V(x, \eta) = p_1 V_1(x_1, \eta_1) + \cdots + p_N V_N(x_N, \eta_N) + \eta^T Q \eta \tag{8.20}
$$

where $p_i > 0$, $i = 1, \ldots, N$, and $Q = Q^T \geq 0$ are decision variables. From (8.16) and (8.17), the derivative of $V(x, \eta)$ along the system equations is upper bounded by

$$
\begin{bmatrix} z_1 \\ \vdots \\ z_N \end{bmatrix}^T \begin{bmatrix} p_1 X_1 & & \\ & \ddots & \\ & & p_N X_N \end{bmatrix} \begin{bmatrix} z_1 \\ \vdots \\ z_N \end{bmatrix} + \begin{bmatrix} \eta \\ y \end{bmatrix}^T \begin{bmatrix} A^T Q + QA & QBS\begin{bmatrix} M \\ I \end{bmatrix} \\ \begin{bmatrix} M \\ I \end{bmatrix}^T S^T B^T Q & 0 \end{bmatrix} \begin{bmatrix} \eta \\ y \end{bmatrix} \tag{8.21}
$$

where, upon substitution of (8.18) for z, the first term becomes

$$
\begin{bmatrix} \eta \\ y \end{bmatrix}^T \begin{bmatrix} C & DS\begin{bmatrix} M \\ I \end{bmatrix} \end{bmatrix}^T \begin{bmatrix} p_1 X_1 & & \\ & \ddots & \\ & & p_N X_N \end{bmatrix} \begin{bmatrix} C & DS\begin{bmatrix} M \\ I \end{bmatrix} \end{bmatrix} \begin{bmatrix} \eta \\ y \end{bmatrix}. \tag{8.22}
$$

Thus, to certify stability, we search for $Q = Q^T \geq 0$ and $p_i > 0$ such that

$$
\begin{bmatrix} A^T Q + QA & QBS\begin{bmatrix} M \\ I \end{bmatrix} \\ \begin{bmatrix} M \\ I \end{bmatrix}^T S^T B^T Q & 0 \end{bmatrix} + \begin{bmatrix} C & DS\begin{bmatrix} M \\ I \end{bmatrix} \end{bmatrix}^T \begin{bmatrix} p_1 X_1 & & \\ & \ddots & \\ & & p_N X_N \end{bmatrix} \begin{bmatrix} C & DS\begin{bmatrix} M \\ I \end{bmatrix} \end{bmatrix} \leq 0. \tag{8.23}
$$

Proposition 8.1 *Consider the interconnected system (2.1)–(2.3) with $f_i(0, 0) = 0$, $h_i(0, 0) = 0$, and suppose each subsystem is dissipative with a positive definite, continuously differentiable storage function $V_i(\cdot, \cdot)$ satisfying (8.16) for some auxiliary system (8.14) and (8.15). If there exist $p_i > 0$, $i = 1, \ldots, N$, and $Q = Q^T \geq 0$ such that (8.23) holds then $x = 0$ is stable.*

This result encompasses Proposition 2.1 as a special case because, when $Q = 0$, $C = 0$, $D = I$, (8.23) becomes

$$\begin{bmatrix} M \\ I \end{bmatrix}^T S^T \begin{bmatrix} p_1 X_1 & & \\ & \ddots & \\ & & p_N X_N \end{bmatrix} S \begin{bmatrix} M \\ I \end{bmatrix} = \begin{bmatrix} M \\ I \end{bmatrix}^T \mathbf{X}(p_1 X_1, \cdots, p_N X_N) \begin{bmatrix} M \\ I \end{bmatrix} \leq 0.$$

Proposition 8.1 infers the stability of $x = 0$ indirectly from the stability of $(x, \eta) = (0, 0)$ for the augmented system where the x subsystem evolves independently and drives the virtual η subsystem. It may appear circuitous to analyze the augmented system rather than search directly for a Lyapunov function $V(x)$. However, the advantage of $V(x, \eta)$ in (8.20) is its separability in x_i which allows for a compositional construction of this function. Indeed the following example shows that a separable Lyapunov function $V(x)$ may not exist when a separable $V(x, \eta)$ as in (8.20) does.

Example 8.3 Suppose system (8.10) in Example 8.2 with $k = 0.5$ is interconnected in negative feedback with the system (8.5) in Example 8.1 with $\alpha = 0.6$ and $\gamma = 6$. Relabeling x in Example 8.1 as x_3, we write the composite system as

$$\frac{dx_1(t)}{dt} = x_2(t)$$

$$\frac{dx_2(t)}{dt} = -x_1(t) - 0.5x_2(t) - 6x_3(t) \tag{8.24}$$

$$\frac{dx_3(t)}{dt} = x_1(t) + x_2(t) - 0.6x_3(t)$$

which, as we show in Appendix C, does not admit a block separable Lyapunov function $V_1(x_1, x_2) + V_2(x_3)$.

In contrast, we here show that a Lyapunov function of the form

$$V_1(x_1, x_2, \eta_1) + V_2(x_3, \eta_2) + \eta^T Q \eta \tag{8.25}$$

exists where η_1 is the state of (8.11) and η_2 is the state of (8.6). Likewise we denote with u_1, y_1, and z_1 the respective variables in Example 8.2 and by u_2, y_2, and z_2 those in Example 8.1, and note that the interconnection matrix is

$$M = \begin{bmatrix} 0 & -1 \\ 1 & 0 \end{bmatrix}.$$

We select $\beta \in (0.5, 0.6)$ so that condition $\beta > 1 - k$ in Example 8.2 and $\beta < \min\{\alpha, 1\}$ in Example 8.1 are satisfied. Thus, there exist quadratic positive definite storage functions $V_1(x_1, x_2, \eta_1)$ and $V_2(x_3, \eta_2)$ satisfying (8.16) with, respectively,

$$X_1 = \begin{bmatrix} 0 & 1/2 \\ 1/2 & 0 \end{bmatrix} \quad \text{and} \quad X_2 = \begin{bmatrix} 0 & 1/2 \\ 1/2 & -\varepsilon \end{bmatrix}, \quad \varepsilon > 0.$$

Next we form the matrices in (8.17) and (8.18):

$$A = \begin{bmatrix} -1 & 0 \\ 0 & -1 \end{bmatrix} \quad BS \begin{bmatrix} M \\ I \end{bmatrix} = \begin{bmatrix} 1 & 0 \\ 1 & 0 \end{bmatrix} \quad C = \begin{bmatrix} 0 & 0 \\ -\beta & 0 \\ 0 & -\beta \\ 0 & 0 \end{bmatrix} \quad DS \begin{bmatrix} M \\ I \end{bmatrix} = \begin{bmatrix} 0 & -1 \\ 1 & 0 \\ 1 & 0 \\ 0 & 1 \end{bmatrix}$$

and check the condition (8.23). It is not difficult to show that (8.23) holds with $p_1 = p_2 = 1$ and

$$Q = q \begin{bmatrix} 1 & -1 \\ -1 & 1 \end{bmatrix} \quad q \geq \frac{\beta^2}{8\varepsilon}$$

thus proving stability with a Lyapunov function of the form (8.25).

8.3   Certification of Performance

Now consider the interconnection in Fig. 5.1 with exogenous input d and performance output e, and introduce a stable linear system

$$\frac{d}{dt}\eta_{N+1}(t) = A_{N+1}\eta_{N+1}(t) + B_{N+1}\begin{bmatrix} d(t) \\ e(t) \end{bmatrix} \quad \eta_{N+1}(t) \in \mathbb{R}^{n'_{N+1}} \quad (8.26)$$

$$z_{N+1}(t) = C_{N+1}\eta_{N+1}(t) + D_{N+1}\begin{bmatrix} d(t) \\ e(t) \end{bmatrix} \quad z_{N+1}(t) \in \mathbb{R}^{p'_{N+1}} \quad (8.27)$$

that serves as a virtual filter for d and e.

The goal is now to certify that the interconnected system is dissipative with respect to the dynamic supply rate

$$z_{N+1}^T W z_{N+1} \quad (8.28)$$

where z_{N+1} is the output of (8.26) and (8.27) and W is a symmetric matrix. The choice of W and $A_{N+1}, B_{N+1}, C_{N+1}, D_{N+1}$ of (8.26) and (8.27) dictate the performance criterion to be certified for the interconnected system.

We assume each subsystem is dissipative with a positive semidefinite, continuously differentiable storage function $V_i(\cdot, \cdot)$ and supply rate $z_i^T X_i z_i$, satisfying (8.16).

We define A, B, C, D to be block diagonal matrices comprised of A_i, B_i, C_i, D_i, $i = 1, \ldots, N + 1$. Similarly to the stability certification, we lump (8.14) and (8.15) and (8.26) and (8.27) into a single auxiliary system

$$\frac{d}{dt}\eta(t) = A\eta(t) + B\overline{S}\begin{bmatrix} u(t) \\ e(t) \\ y(t) \\ d(t) \end{bmatrix} = A\eta(t) + B\overline{S}\begin{bmatrix} \overline{M} \\ I \end{bmatrix}\begin{bmatrix} y(t) \\ d(t) \end{bmatrix} \tag{8.29}$$

$$z(t) = C\eta(t) + D\overline{S}\begin{bmatrix} u(t) \\ e(t) \\ y(t) \\ d(t) \end{bmatrix} = C\eta(t) + D\overline{S}\begin{bmatrix} \overline{M} \\ I \end{bmatrix}\begin{bmatrix} y(t) \\ d(t) \end{bmatrix} \tag{8.30}$$

where $\overline{M}$ is the interconnection matrix (5.1) and $\overline{S}$ is a permutation matrix such that

$$\overline{S}\begin{bmatrix} u_1 \\ \vdots \\ u_N \\ e \\ y_1 \\ \vdots \\ y_N \\ d \end{bmatrix} = \begin{bmatrix} u_1 \\ y_1 \\ \vdots \\ u_N \\ y_N \\ d \\ e \end{bmatrix}. \tag{8.31}$$

Next we search for a storage function of the form (8.20) where $p_i \geq 0$, $i = 1, \ldots, N$ and $Q = Q^T \geq 0$ are decision variables. The derivative of $V(x, \eta)$ along the system equations is upper bounded by the supply rate $z_{N+1}^T W z_{N+1}$ if

$$\begin{bmatrix} A^T Q + QA & QB\overline{S}\begin{bmatrix} \overline{M} \\ I \end{bmatrix} \\ \begin{bmatrix} \overline{M} \\ I \end{bmatrix}^T \overline{S}^T B^T Q & 0 \end{bmatrix} + \begin{bmatrix} C & D\overline{S}\begin{bmatrix} \overline{M} \\ I \end{bmatrix} \end{bmatrix}^T \begin{bmatrix} p_1 X_1 & & \\ & \ddots & \\ & & p_N X_N \\ & & & -W \end{bmatrix} \begin{bmatrix} C & D\overline{S}\begin{bmatrix} \overline{M} \\ I \end{bmatrix} \end{bmatrix} \leq 0. \tag{8.32}$$

Proposition 8.2 *Consider the subsystems (2.1) and (2.2) with $f_i(0, 0) = 0$, $h_i(0, 0) = 0$ interconnected by (5.1). Suppose each subsystem is dissipative with a positive semidefinite, continuously differentiable storage function $V_i(\cdot, \cdot)$ satisfying (8.16) for some auxiliary system (8.14) and (8.15). If there exist $p_i \geq 0$, $i = 1, \ldots, N$, and $Q = Q^T \geq 0$ such that (8.32) holds then the system is dissipative with respect to the dynamic supply rate (8.28).*

8.4 Search for Dynamic Supply Rates

In Sect. 6.2 the ADMM algorithm was used to search for feasible subsystem dissipativity properties certifying stability or performance. We can also use this method when the subsystem properties are described by dynamic supply rates [1].

For each subsystem the auxiliary system (8.14) and (8.15) is fixed and the matrices $X_1, \ldots, X_N$ in (8.23) or (8.32) are decision variables where each X_i must satisfy the *local constraint* (8.16). Since each X_i is a decision variable we can drop the scaling weights p_i from (8.23) and (8.32). Thus, for performance certification the *global constraint* becomes

$$
\begin{bmatrix} A^T Q + QA & QB\overline{S}\begin{bmatrix} \overline{M} \\ I \end{bmatrix} \\ \begin{bmatrix} \overline{M} \\ I \end{bmatrix}^T \overline{S}^T B^T Q & 0 \end{bmatrix} + \begin{bmatrix} C & D\overline{S}\begin{bmatrix} \overline{M} \\ I \end{bmatrix} \end{bmatrix}^T \begin{bmatrix} X_1 & & \\ & \ddots & \\ & & X_N \\ & & & -W \end{bmatrix} \begin{bmatrix} C & D\overline{S}\begin{bmatrix} \overline{M} \\ I \end{bmatrix} \end{bmatrix} \leq 0
$$

$$(8.33)$$

and the ADMM algorithm takes the following form.

X-**updates:** For each i, solve the local problem

$$
X_i^{k+1} = \arg\min_{X \text{ s.t. } (8.16) \text{ with } V \geq 0} \left\| X - Z_i^k + S_i^k \right\|_F^2
$$

where $\|\cdot\|_F$ represents the Frobenius norm.

Z-**update:** If $X_1^{k+1}, \ldots, X_N^{k+1}$ satisfy (8.33), then terminate. Otherwise, solve the global problem

$$
Z_{1:N}^{k+1} = \arg\min_{(Z_1, \ldots, Z_N) \text{ s.t. } (8.33)} \sum_{i=1}^{N} \left\| X_i^{k+1} - Z_i + S_i^k \right\|_F^2 .
$$

S-**updates:** Update S_i by

$$
S_i^{k+1} = X_i^{k+1} - Z_i^{k+1} + S_i^k
$$

and return to the *X*-updates.

For stability certification we replace (8.33) by (8.23), again with the weights p_i dropped. An extension of the symmetry reduction techniques in Chap. 7 to dynamic supply rates is pursued in [2].

8.5 EID with Dynamic Supply Rates

Consider the system (3.3) and (3.4) and suppose there exists a set $\mathscr{X} \subset \mathbb{R}^n$ where, for every $\bar{x} \in \mathscr{X}$, there exists unique $\bar{u} \in \mathbb{R}^m$ satisfying $f(\bar{x}, \bar{u}) = 0$. We append to this system the stable linear system (8.1) and (8.2) where all eigenvalues of A have negative real parts. Thus A is invertible and there exists a unique $\bar{\eta}$ such that

$$A\bar{\eta} + B \begin{bmatrix} \bar{u} \\ \bar{y} \end{bmatrix} = 0 \tag{8.34}$$

where $\bar{y} \triangleq h(\bar{x}, \bar{u})$. Likewise we define

$$\bar{z} = C\bar{\eta} + D \begin{bmatrix} \bar{u} \\ \bar{y} \end{bmatrix}, \tag{8.35}$$

and note that $\bar{u}$, $\bar{y}$, $\bar{\eta}$, and $\bar{z}$ are functions of $\bar{x}$.

Definition 8.2 We say that the system (3.3) and (3.4) is **equilibrium independent dissipative (EID)** with the dynamic supply rate $z^T X z$ where z is the output of (8.1) and (8.2) and X is a real symmetric matrix if there exists a storage function $V : \mathbb{R}^n \times \mathbb{R}^{n'} \times \mathscr{X} \times \mathbb{R}^{n'} \mapsto \mathbb{R}$ such that $V(\bar{x}, \bar{\eta}, \bar{x}, \bar{\eta}) = 0$, $V(x, \eta, \bar{x}, \bar{\eta}) \geq 0$ for all $(x, \eta, \bar{x}, \bar{\eta}) \in \mathbb{R}^n \times \mathbb{R}^{n'} \times \mathscr{X} \times \mathbb{R}^{n'}$, and

$$\nabla_x V(x, \eta, \bar{x}, \bar{\eta})^T f(x, u) + \nabla_\eta V(x, \eta, \bar{x}, \bar{\eta})^T \left(A\eta + B \begin{bmatrix} u \\ y \end{bmatrix} \right) \leq (z - \bar{z})^T X (z - \bar{z}) \tag{8.36}$$

for all $(x, \eta, \bar{x}, u) \in \mathbb{R}^n \times \mathbb{R}^{n'} \times \mathscr{X} \times \mathbb{R}^m$ where $\bar{\eta}$, $\bar{z}$ are as in (8.34) and (8.35).

Propositions (8.1) and (8.2) can be easily generalized to interconnections of EID systems with dynamic supply rates. In this case the stability (8.23) and performance (8.32) criteria are the same, but guarantee negativity of a quadratic inequality in the shifted equilibrium points as in (5.11). Furthermore, the ADMM algorithm can be used by modifying the X-updates to certify EID with respect to a dynamic supply rate for each subsystem.

References

1. Meissen, C., Lessard, L., Arcak, M., Packard, A.K.: Compositional performance certification of interconnected systems using ADMM. Automatica **61**, 55–63 (2015). http://dx.doi.org/10.1016/j.automatica.2015.07.027
2. Rufino Ferreira, A., Meissen, C., Arcak, M., Packard, A.: Symmetry Reduction for Performance Certification of Interconnected Systems (submitted)

Chapter 9
Comparison to Other Input/Output Approaches

Throughout the book we employed a state-space approach with the help of the dissipativity concept, generalized in Chap. 8 to dynamic supply rates. In this final chapter, we make connections to other input/output approaches that treat dynamical systems as operators mapping inputs to outputs in function spaces. We start with the classical techniques summarized in [1, 2], and next relate the dynamic supply rates of Chap. 8 to integral quadratic constraints (IQCs) introduced in [3]. We conclude by pointing to further results that are complementary to those presented in the book.

9.1 The Classical Input/Output Theory

Consider a dynamical system where inputs $u(\cdot)$, assumed to have the property that $\int_0^\tau |u(t)|^2 dt$ is finite for all $\tau \geq 0$, generate outputs $y(\cdot)$ satisfying

$$\int_0^\tau \begin{bmatrix} u(t) \\ y(t) \end{bmatrix}^T X \begin{bmatrix} u(t) \\ y(t) \end{bmatrix} dt \geq 0 \quad \forall \tau \geq 0. \tag{9.1}$$

Note that this property follows from dissipativity (Definition 1.1) with supply rate

$$s(u, y) = \begin{bmatrix} u \\ y \end{bmatrix}^T X \begin{bmatrix} u \\ y \end{bmatrix}$$

when $x(0) = 0$. However, in this section we do not make explicit use of a state model and, thus, do not rely on a storage function. Instead we take (9.1) as a stand-alone property as in the classical input/output approach [1], extended to large-scale interconnections in [2].

© The Author(s) 2016
M. Arcak et al., *Networks of Dissipative Systems*,
SpringerBriefs in Control, Automation and Robotics,
DOI 10.1007/978-3-319-29928-0_9

Now consider the interconnection in Fig. 5.1 with exogenous input d and performance output e, and suppose each subsystem, $i = 1, \ldots, N$, satisfies

$$\int_0^\tau \begin{bmatrix} u_i(t) \\ y_i(t) \end{bmatrix}^T X_i \begin{bmatrix} u_i(t) \\ y_i(t) \end{bmatrix} dt \geq 0 \quad \forall \tau \geq 0. \tag{9.2}$$

Assuming that $\int_0^\tau |d(t)|^2 dt$ is finite for all $\tau \geq 0$ and that the interconnection admits a solution for all $t \geq 0$, we derive an analog of Proposition 5.1 for performance certification without relying on storage functions.

Recall that the main condition of Proposition 5.1 was

$$\begin{bmatrix} M_{uy} & M_{ud} \\ I & 0 \\ 0 & I \\ M_{ey} & M_{ed} \end{bmatrix}^T \begin{bmatrix} \mathbf{X}(p_1 X_1, \ldots, p_N X_N) & 0 \\ 0 & -W \end{bmatrix} \begin{bmatrix} M_{uy} & M_{ud} \\ I & 0 \\ 0 & I \\ M_{ey} & M_{ed} \end{bmatrix} \leq 0, \tag{9.3}$$

which guaranteed

$$\begin{bmatrix} u \\ y \\ d \\ e \end{bmatrix}^T \begin{bmatrix} \mathbf{X}(p_1 X_1, \ldots, p_N X_N) & 0 \\ 0 & -W \end{bmatrix} \begin{bmatrix} u \\ y \\ d \\ e \end{bmatrix} \leq 0. \tag{9.4}$$

It follows from this inequality that

$$\int_0^\tau \begin{bmatrix} d(t) \\ e(t) \end{bmatrix}^T W \begin{bmatrix} d(t) \\ e(t) \end{bmatrix} dt \geq \int_0^\tau \begin{bmatrix} u(t) \\ y(t) \end{bmatrix}^T \mathbf{X}(p_1 X_1, \ldots, p_N X_N) \begin{bmatrix} u(t) \\ y(t) \end{bmatrix} dt$$

$$= \int_0^\tau \left\{ \sum_{i=1}^N p_i \begin{bmatrix} u_i(t) \\ y_i(t) \end{bmatrix}^T X_i \begin{bmatrix} u_i(t) \\ y_i(t) \end{bmatrix} \right\} dt. \tag{9.5}$$

Since $p_i \geq 0$, we conclude from (9.2) that (9.5) is nonnegative; that is,

$$\int_0^\tau \begin{bmatrix} d(t) \\ e(t) \end{bmatrix}^T W \begin{bmatrix} d(t) \\ e(t) \end{bmatrix} dt \geq 0 \quad \forall \tau \geq 0, \tag{9.6}$$

establishing the desired performance property of the interconnection.

In the absence of a state model Lyapunov stability concepts are not applicable; therefore, a direct analog of Proposition 2.1 is not possible. However, when condition (2.8) of this proposition holds with strict inequality that is

$$\begin{bmatrix} M_{uy} \\ I \end{bmatrix}^T \mathbf{X}(p_1 X_1, \ldots, p_N X_N) \begin{bmatrix} M_{uy} \\ I \end{bmatrix} < 0, \tag{9.7}$$

an L_2 stability property is guaranteed where $d(\cdot)$ being an L_2 signal ($\int_0^\infty |d(t)|^2 dt < \infty$) guarantees $e(\cdot)$ to be L_2 as well. To see this let

$$W = \begin{bmatrix} \gamma^2 I & 0 \\ 0 & -I \end{bmatrix} \tag{9.8}$$

and note that the upper left, upper right, and lower right blocks of (9.3) are

$$\Lambda_{11} \triangleq \begin{bmatrix} M_{uy} \\ I \end{bmatrix}^T \mathbf{X}(p_1 X_1, \ldots, p_N X_N) \begin{bmatrix} M_{uy} \\ I \end{bmatrix} + M_{ey}^T M_{ey} \tag{9.9}$$

$$\Lambda_{12} \triangleq \begin{bmatrix} M_{uy} \\ I \end{bmatrix}^T \mathbf{X}(p_1 X_1, \ldots, p_N X_N) \begin{bmatrix} M_{ud} \\ 0 \end{bmatrix} + M_{ey}^T M_{ed} \tag{9.10}$$

$$\Lambda_{22} \triangleq \begin{bmatrix} M_{ud} \\ 0 \end{bmatrix}^T \mathbf{X}(p_1 X_1, \ldots, p_N X_N) \begin{bmatrix} M_{ud} \\ 0 \end{bmatrix} + M_{ed}^T M_{ed} - \gamma^2 I. \tag{9.11}$$

If (9.7) holds, we can scale all coefficients p_i by a sufficiently large constant to dominate $M_{ey}^T M_{ey}$ and ensure $\Lambda_{11} < 0$. Next, we select $\gamma > 0$ large enough to guarantee the Schur complement of Λ_{11}, given by $\Lambda_{22} - \Lambda_{12}^T \Lambda_{11}^{-1} \Lambda_{12}$, is negative definite. This means that $\Lambda < 0$, that is (9.6) holds with (9.8), proving that a finite L_2 gain exists from d to e.

Note that the L_2 stability condition (9.7) does not restrict the matrices M_{ey}, M_{ed}, M_{ud}. In particular the choice $M_{ey} = I$, that is $e = y$, shows that the output of each subsystem is L_2 when $d(\cdot)$ is L_2.

Unlike the pure input/output arguments above, in this book we took a state-space approach that allowed us to account for initial conditions, to establish Lyapunov stability and safety properties using bounds on the storage functions, and to develop criteria that do not depend on the exact knowledge of the network equilibrium.

9.2 Integral Quadratic Constraints (IQCs)

In this section, we relate dynamic supply rates (Chap. 8) to the frequency domain notion of *integral quadratic constraints* [3].

Definition 9.1 Let $\hat{u}$ denote the Fourier transform of $u \in L_2^m$ and let Π : $\mathbb{R} \rightarrow \mathbb{C}^{(m+p)\times(m+p)}$ be a measurable, bounded, Hermitian-valued function. A bounded, causal operator G mapping L_2^m to L_2^p is said to satisfy the integral quadratic constraint (IQC) defined by Π if for all $u \in L_2^m$, $y = Gu$ satisfies

$$\int_{-\infty}^{\infty} \begin{bmatrix} \hat{u}(\omega) \\ \hat{y}(\omega) \end{bmatrix}^* \Pi(\omega) \begin{bmatrix} \hat{u}(\omega) \\ \hat{y}(\omega) \end{bmatrix} d\omega \geq 0. \qquad (9.12)$$

The time domain constraint (9.1) with $\tau = \infty$ implies the IQC defined by $\Pi = X$ because, from Parseval's Theorem (see, e.g., [1, Theorem B.2.4]),

$$\int_0^{\infty} \begin{bmatrix} u(t) \\ y(t) \end{bmatrix}^T X \begin{bmatrix} u(t) \\ y(t) \end{bmatrix} dt = \frac{1}{2\pi} \int_{-\infty}^{\infty} \begin{bmatrix} \hat{u}(\omega) \\ \hat{y}(\omega) \end{bmatrix}^* X \begin{bmatrix} \hat{u}(\omega) \\ \hat{y}(\omega) \end{bmatrix} d\omega \geq 0. \qquad (9.13)$$

Likewise, (8.3) with $x(0) = 0$, $\eta(0) = 0$, and $\tau = \infty$ implies

$$\int_0^{\infty} z(t)^T X z(t) dt = \frac{1}{2\pi} \int_{-\infty}^{\infty} \hat{z}(\omega)^* X \hat{z}(\omega) d\omega \geq 0. \qquad (9.14)$$

Substituting into (9.14)

$$\hat{z}(\omega) = \Psi(\omega) \begin{bmatrix} \hat{u}(\omega) \\ \hat{y}(\omega) \end{bmatrix}, \qquad (9.15)$$

which follows from (8.1)–(8.2) with $\Psi(\omega) = D + C(j\omega I - A)^{-1} B$, we obtain

$$\int_{-\infty}^{\infty} \begin{bmatrix} \hat{u}(\omega) \\ \hat{y}(\omega) \end{bmatrix}^* \underbrace{\Psi(\omega)^* X \Psi(\omega)}_{= \Pi(\omega)} \begin{bmatrix} \hat{u}(\omega) \\ \hat{y}(\omega) \end{bmatrix} d\omega \geq 0. \qquad (9.16)$$

Thus, the dynamic supply rate in Definition 8.1 leads to an IQC with $\Pi(\omega) = \Psi(\omega)^* X \Psi(\omega)$ where $\Psi(\omega)$ is dictated by the filter (8.1)–(8.2).

Next, consider the concatenation of N subsystems as in Fig. 9.1 where each subsystem G_i with input u_i and output y_i satisfies an IQC defined by Π_i. Then, for any set of coefficients $p_i \geq 0$, we have

$$\int_{-\infty}^{\infty} \begin{bmatrix} \hat{u}(\omega) \\ \hat{y}(\omega) \end{bmatrix}^* S^T \begin{bmatrix} p_1 \Pi_1(\omega) & & \\ & \ddots & \\ & & p_N \Pi_N(\omega) \end{bmatrix} S \begin{bmatrix} \hat{u}(\omega) \\ \hat{y}(\omega) \end{bmatrix} d\omega \geq 0 \qquad (9.17)$$

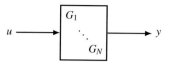

Fig. 9.1 Concatenation of subsystems $G_1, \ldots, G_N$ where $u = [u_1^T \ldots u_N^T]^T$ and $y = [y_1^T \ldots y_N^T]^T$. If each subsystem G_i with input u_i and output y_i satisfies an IQC defined by Π_i, then the combined system satisfies the IQC defined by (9.18) for any set of coefficients $p_i \geq 0$

where S is the permutation matrix defined in (8.19). Thus, the combined system satisfies the IQC defined by

$$\Pi(\omega) = S^T \begin{bmatrix} p_1 \Pi_1(\omega) & & \\ & \ddots & \\ & & p_N \Pi_N(\omega) \end{bmatrix} S = \mathbf{X}(p_1 \Pi_1(\omega), \ldots, p_N \Pi_N(\omega)).$$

(9.18)

9.3 The IQC Stability Theorem

We now return to the interconnection in Fig. 2.1 and relate the stability criterion (8.23) to the frequency domain inequality

$$\begin{bmatrix} M \\ I \end{bmatrix}^T \mathbf{X}(p_1 \Pi_1(\omega), \ldots, p_N \Pi_N(\omega)) \begin{bmatrix} M \\ I \end{bmatrix} \leq 0 \quad \forall \omega \in \mathbb{R}$$

(9.19)

$$\Pi_i(\omega) = \Psi_i(\omega)^* X_i \Psi(\omega) \quad \Psi_i(\omega) = D_i + C_i(j\omega I - A_i)^{-1} B_i.$$

To this end we use (9.18) and rewrite the matrix in (9.19) as

$$\begin{bmatrix} M \\ I \end{bmatrix}^T S^T \begin{bmatrix} p_1 \Pi_1(\omega) & & \\ & \ddots & \\ & & p_N \Pi_N(\omega) \end{bmatrix} S \begin{bmatrix} M \\ I \end{bmatrix}$$

$$= \begin{bmatrix} M \\ I \end{bmatrix}^T S^T \Psi(\omega)^* \begin{bmatrix} p_1 X_1 & & \\ & \ddots & \\ & & p_N X_N \end{bmatrix} \Psi(\omega) S \begin{bmatrix} M \\ I \end{bmatrix}$$

(9.20)

where

$$\Psi(\omega) = \begin{bmatrix} \Psi_1(\omega) & & \\ & \ddots & \\ & & \Psi_N(\omega) \end{bmatrix} = D + C(j\omega I - A)^{-1} B$$

(9.21)

and A, B, C, D are block diagonal matrices comprised of A_i, B_i, C_i. D_i, $i = 1, \ldots, N$. Defining

$$\overline{B} \triangleq BS \begin{bmatrix} M \\ I \end{bmatrix} \quad \overline{D} \triangleq DS \begin{bmatrix} M \\ I \end{bmatrix} \tag{9.22}$$

and substituting

$$\Psi(\omega)S \begin{bmatrix} M \\ I \end{bmatrix} = \overline{D} + C(j\omega I - A)^{-1}\overline{B} \tag{9.23}$$

in (9.20), we rewrite (9.19) as

$$\begin{bmatrix} (j\omega I - A)^{-1}\overline{B} \\ I \end{bmatrix}^* [C \ \overline{D}]^T \begin{bmatrix} p_1 X_1 \\ & \ddots \\ & & p_N X_N \end{bmatrix} [C \ \overline{D}] \begin{bmatrix} (j\omega I - A)^{-1}\overline{B} \\ I \end{bmatrix} \leq 0. \tag{9.24}$$

When A is Hurwitz and $(A, \overline{B})$ is controllable, Theorem C.1 in Appendix C states that (9.24) is equivalent to the existence of $Q = Q^T$ such that

$$\begin{bmatrix} A^T Q + QA & Q\overline{B} \\ \overline{B}^T Q & 0 \end{bmatrix} + [C \ \overline{D}]^T \begin{bmatrix} p_1 X_1 \\ & \ddots \\ & & p_N X_N \end{bmatrix} [C \ \overline{D}] \leq 0 \tag{9.25}$$

which is identical to (8.23). In particular, $Q \geq 0$ when the upper left block of the second term on the left-hand side is positive semidefinite.

A similar derivation relates the performance criterion (8.32) for the interconnection in Fig. 5.1 to the frequency domain condition

$$\begin{bmatrix} M_{uy} & M_{ud} \\ I & 0 \\ 0 & I \\ M_{ey} & M_{ed} \end{bmatrix}^T \begin{bmatrix} \mathbf{X}(p_1\Pi_1(\omega), \ldots, p_N\Pi_N(\omega)) & 0 \\ 0 & -\Pi_W(\omega) \end{bmatrix} \begin{bmatrix} M_{uy} & M_{ud} \\ I & 0 \\ 0 & I \\ M_{ey} & M_{ed} \end{bmatrix} \leq 0 \quad \forall \omega \in \mathbb{R} \tag{9.26}$$

where $\Pi_W(\omega)$ is obtained from the performance supply rate (8.26)–(8.28) by

$$\Pi_W(\omega) = \Psi_{N+1}(\omega)^* W \Psi_{N+1}(\omega) \quad \Psi_{N+1}(\omega) = D_{N+1} + C_{N+1}(j\omega I - A_{N+1})^{-1} B_{N+1}.$$

For the finite L_2 gain supply rate $\Pi_W(\omega) = W$ given in (9.8), arguments similar to those in Sect. 9.1 show that (9.26) holds for sufficiently large γ if, for some $\mu > 0$,

$$\begin{bmatrix} M_{uy} \\ I \end{bmatrix}^T \mathbf{X}(p_1\Pi_1(\omega), \ldots, p_N\Pi_N(\omega)) \begin{bmatrix} M_{uy} \\ I \end{bmatrix} \leq -\mu I \quad \forall \omega \in \mathbb{R}. \tag{9.27}$$

Indeed (9.27) is the main condition of the IQC Stability Theorem [3], when adapted to the interconnection in Fig. 5.1:

Theorem 9.1 *Suppose each G_i is a bounded, causal operator mapping $L_2^{m_i}$ to $L_2^{p_i}$ such that, for every $\kappa \in [0, 1]$, the interconnection of κG_i as in Fig. 5.1 is well posed and κG_i satisfies the IQC defined by Π_i, $i = 1, \ldots, N$. Under these conditions, if there exist $p_i \geq 0$ and $\mu > 0$ satisfying (9.27) then the interconnection for $\kappa = 1$ is L_2 stable.*

Although the KYP Lemma (Appendix C) relates frequency domain inequalities such as (9.19), (9.26) and (9.27) above to LMIs derived with the dissipativity approach, several technical discrepancies exist between the IQC and dissipativity approaches. First, the KYP Lemma does not guarantee a positive semidefinite solution to the LMI (C.2) whereas semidefiniteness is required in the dissipativity approach. Second, from Parseval's Theorem, the frequency domain IQC definition (9.12) is equivalent to (9.1) with $\tau = \infty$ which is less restrictive than dissipativity which implies (9.1) for all $\tau \geq 0$.

On the other hand, the IQC stability theorem quoted above relies on the extra assumption that the scaled operators κG_i satisfy the IQC defined by Π_i and that their interconnection remain well posed for $\kappa \in [0, 1]$. Reconciling the IQC and dissipativity approaches is an active research topic, with partial results reported in [4, 5] and the references therein.

9.4 Conclusions and Further Results

In this book, we presented a compositional approach to certify desirable properties of an interconnection from dissipativity characteristics of the subsystems. Despite its computational benefits, however, this bottom-up approach may introduce conservatism and understanding the extent of such conservatism is an important topic for further study.

In [6, Theorem 3] we showed that certifying stability and performance of a linear system from dissipativity of its subsystems is no more conservative than searching for separable Lyapunov and storage functions. In Example 8.3 of this book we showed that, by augmenting the dynamics of the subsystems with appropriate filters (i.e., by using dynamic supply rates) we may be able to find separable Lyapunov functions in situations where no separable Lyapunov function exists without such filters. Further connections to separable Lyapunov and storage functions would enable a unified perspective for compositional system analysis.

In this book, we primarily employed quadratic supply rates, such as those for passivity and finite L_2 gain properties. Another commonly used dissipativity property is *input to state stability* (ISS) [7] which has been used to derive ISS small gain theorems in [8, 9], extended to large-scale interconnections in [10, 11].

A common concern when stability certificates are derived from dissipativity is robustness against sampling and time delays. The degradation of dissipativity under sampling is studied in [12] and the results can be adapted to the interconnections in this book. For robustness against time delays, [13] employed a variant of the IQC stability theorem above. This paper first notes that dissipativity with a static supply rate does not encapsulate time scale information (see Appendix D, Problem 9), disallowing stability estimates where the effect of delay depends on its duration relative to the time scales of the dynamics. To overcome this problem, it introduces a complementary "roll off" IQC that is frequency dependent and provides the missing time scale information. It then derives a stability condition that degrades gracefully with the duration of delay.

The dissipativity approach to networks in this book was partially motivated by multi-agent systems where bidirectional communication yields a skew symmetric interconnection, as illustrated in Sect. 4.2. The compatibility of this structure with passivity properties was fully harnessed in [14] to derive distributed and adaptive control techniques. Synchronization problems that arise in multi-agent systems and numerous other networks was studied with a related input/output approach in [15].

We restricted our attention to dissipativity properties that are global in the state and input spaces. Local variants and corresponding computational procedures have been pursued in [16, 17]. Finally, a stochastic stability test was developed in [18] that extends the compositional methods in Chaps. 2 and 3 to stochastic differential equations.

References

1. Desoer, C., Vidyasagar, M.: Feedback systems: input-output properties. Society for industrial and applied mathematics, Philadelphia (2009). Academic Press, New York (1975)
2. Vidyasagar, M.: Input-Output Analysis of Large Scale Interconnected Systems. Springer, Berlin (1981)
3. Megretski, A., Rantzer, A.: System analysis via integral quadratic constraints. IEEE Trans. Autom. Control **42**, 819–830 (1997)
4. Veenman, J., Scherer, C.: Stability analysis with integral quadratic constraints: a dissipativity based proof. In: IEEE 52nd Annual Conference on Decision and Control (CDC), pp. 3770–3775 (2013).10.1109/CDC.2013.6760464
5. Seiler, P.: Stability analysis with dissipation inequalities and integral quadratic constraints. IEEE Trans. Autom. Control **60**(6), 1704–1709 (2015).10.1109/TAC.2014.2361004
6. Meissen, C., Lessard, L., Arcak, M., Packard, A.K.: Compositional performance certification of interconnected systems using ADMM. Automatica **61**, 55–63 (2015). http://dx.doi.org/10.1016/j.automatica.2015.07.027
7. Sontag, E.: Smooth stabilization implies coprime factorization. IEEE Trans. Autom. Control **34**, 435–443 (1989)

8. Jiang, Z.P., Teel, A., Praly, L.: Small-gain theorem for ISS systems and applications. Math. Control Signals Systems **7**, 95–120 (1994)
9. Teel, A.: A nonlinear small gain theorem for the analysis of control systems with saturation. IEEE Trans. Autom. Control **41**(9), 1256–1271 (1996)
10. Dashkovskiy, S., Ruffer, B., Wirth, F.: An ISS small-gain theorem for general networks. Math. Control Signals Systems **19**, 93–122 (2007)
11. Dashkovskiy, S., Ruffer, B., Wirth, F.: Small gain theorems for large scale systems and construction of ISS Lyapunov functions. SIAM J. Control Optim. **48**, 4089–4118 (2010)
12. Laila, D., Nešić, D., Teel, A.: Open and closed loop dissipation inequalities under sampling and controller emulation. Eur. J. Control **8**(2), 109–125 (2002)
13. Summers, E., Arcak, M., Packard, A.: Delay robustness of interconnected passive systems: an integral quadratic constraint approach. IEEE Trans. Autom. Control **58**(3), 712–724 (2013)
14. Bai, H., Arcak, M., Wen, J.: Cooperative Control Design: A Systematic, Passivity-Based Approach. Communications and Control Engineering. Springer, New York (2011)
15. Scardovi, L., Arcak, M., Sontag, E.: Synchronization of interconnected systems with applications to biochemical networks: an input-output approach. IEEE Trans. Autom. Control **55**(6), 1367–1379 (2010)
16. Topcu, U., Packard, A., Seiler, P.: Local stability analysis using simulations and sum-of-squares programming. Automatica **44**(10), 2669–2675 (2008)
17. Summers, E., Chakraborty, A., Tan, W., Topcu, U., Seiler, P., Balas, G., Packard, A.: Quantitative local L2-gain and reachability analysis for nonlinear systems. Int. J. Robust Nonlinear Control **23**(10), 1115–1135 (2013)
18. Ferreira, A., Arcak, M., Sontag, E.: Stability certification of large scale stochastic systems using dissipativity. Automatica **48**(11), 2956–2964 (2012)

Appendix A
Sum-of-Squares (SOS) Programming

Many of the algebraic conditions derived in this book involve an expression that must be nonnegative for all values of the independent variables. For example, dissipativity requires $s(u, h(x, u)) - \nabla V(x)^T f(x, u) \geq 0$ and $V(x) \geq 0$ for all values of x and u. Checking this nonnegativity for given $\{f, h, s, V\}$ can be challenging. In the special case that f and h are linear and V and s quadratic, the nonnegativity conditions are simple matrix semidefinite constraints, where the matrices in question are affine functions of the quadratic forms that define V and s. When these functions are more general polynomials, other computational tools are needed.

In its basic form, *SOS programming* is a computationally viable way to verify that real multivariable polynomials are nonnegative. Recall that a *monomial* is a product of powers of variables with nonnegative integer exponents, for example, $m(x) := x_1^2 x_2$. The *degree* of a monomial is the sum of its exponents, so the degree of m is 3. A *polynomial* is a finite linear combination of monomials, for example,

$$q(x_1, x_2) \triangleq x_1^2 - 2x_1 x_2^2 + 2x_1^4 + 2x_1^3 x_2 - x_1^2 x_2^2 + 6x_2^4. \tag{A.1}$$

Let $\mathbb{R}[x]$ denote the set of all polynomials in variables $x \in \mathbb{R}^n$, and let θ denote the identically zero polynomial. The *degree* of a polynomial p, denoted $\partial(p)$, is the maximum degree of its monomials. In (A.1) above, $\partial(q) = 4$.

Definition A.1 A polynomial p is a **sum-of-squares** (SOS) if there exists polynomials $g_1, \ldots, g_N$ such that $p = \sum_{i=1}^{N} g_i^2$.

Within the set of all polynomials $\mathbb{R}[x]$, let $\Sigma[x]$ denote the set of all SOS polynomials. One trivial but important fact is if $p \in \Sigma$, then p is nonnegative everywhere, since its value is the sum-of-squares of values of other polynomials.

© The Author(s) 2016
M. Arcak et al., *Networks of Dissipative Systems*,
SpringerBriefs in Control, Automation and Robotics,
DOI 10.1007/978-3-319-29928-0

The polynomial $q(x_1, x_2)$ in (A.1) is a SOS because it can be expressed as

$$q(x_1, x_2) = (x_1 - x_2^2)^2 + \frac{1}{2}\left(2x_1^2 - 3x_2^2 + x_1 x_2\right)^2 + \frac{1}{2}\left(x_2^2 + 3x_1 x_2\right)^2.$$

This equality is easy to verify: simply multiply out and match terms. What is less clear is how this decomposition was obtained. Semidefinite programming can ascertain such decompositions, or determine that none is possible.

Let $z(x)$ be the vector of all monomials in n variables, of degree $\leq d$,

$$z(x) \triangleq [1, x_1, x_2, \ldots, x_n, x_1^2, x_1 x_2, \ldots, x_n^d]^T.$$

Obviously z depends on n and d, but the additional notation is suppressed for clarity. The length of z is

$$l_{[n,d]} \triangleq \binom{n+d}{d}.$$

For any polynomial p with $\partial(p) \leq d$, there is a unique $c \in \mathbb{R}^{l_{[n,d]}}$ such that $p = c^T z$, moreover c depends linearly on p. Clearly, c contains the coefficients of the monomials in the summation that makes up p.

Other representations of p are possible. Taking *all* products of any two elements of z gives all (with some repetitions) monomials of degree $\leq 2d$. This leads to the *Gram matrix representation.*

Definition A.2 For every polynomial p with $\partial(p) \leq 2d$, there is a symmetric matrix $Q \in \mathbb{R}^{l_{[n,d]} \times l_{[n,d]}}$ such that $p(x) = z(x)^T Q z(x)$. This is called a **Gram matrix representation** of p.

The Gram matrix representation is not unique. For example, take $n = d = 2$ so that

$$z(x) \triangleq [1, x_1, x_2, x_1^2, x_1 x_2, x_2^2]^T.$$

With $p \triangleq 4x_1^2 x_2^2$, both

$$Q_1 = \begin{bmatrix} 0 & 0 & 0 & 0 & 0 & 0 \\ 0 & 0 & 0 & 0 & 0 & 0 \\ 0 & 0 & 0 & 0 & 0 & 0 \\ 0 & 0 & 0 & 0 & 0 & 0 \\ 0 & 0 & 0 & 0 & 4 & 0 \\ 0 & 0 & 0 & 0 & 0 & 0 \end{bmatrix}, \quad Q_2 = \begin{bmatrix} 0 & 0 & 0 & 0 & 0 & 0 \\ 0 & 0 & 0 & 0 & 0 & 0 \\ 0 & 0 & 0 & 0 & 0 & 0 \\ 0 & 0 & 0 & 0 & 0 & 2 \\ 0 & 0 & 0 & 0 & 0 & 0 \\ 0 & 0 & 0 & 2 & 0 & 0 \end{bmatrix}$$

give $p = z^T Q_i z$. Nevertheless, Gram matrix representations of polynomials play a key role in the sum-of-squares decomposition [1, 2].

> **Theorem A.1** *A polynomial p with $\partial(p) \leq 2d$ is SOS if and only if there exists $Q = Q^T \succeq 0$ such that $p(x) = z(x)^T Q z(x)$ for all $x \in \mathbb{R}^n$, where $z(x)$ is the vector of all monomials of degree up to d.*

Proof It is easy to see that

$$p \text{ is SOS} \Leftrightarrow \exists \text{ polynomials } \{g_i\}_{i=1}^N \text{ such that } p = \sum_{i=1}^N g_i^2$$

$$\Leftrightarrow \exists \text{ vectors } \{L_i\}_{i=1}^N \subset \mathbb{R}^{l_{[n,d]}} \text{ such that } p = \sum_{i=1}^N (L_i z)^2$$

$$\Leftrightarrow \exists \text{ a matrix } L \in \mathbb{R}^{N \times l_{[n,d]}} \text{ such that } p = z^T L^T L z$$

$$\Leftrightarrow \exists \text{ a matrix } Q \succeq 0 \text{ such that } p = z^T Q z.$$

In the example, $p = (2x_1x_2)^2$ is a sum-of squares and $Q_1 \succeq 0$ (confirming the claim of Theorem A.1), but Q_2 is indefinite (illustrating that not all Q satisfying $p = z^T Q z$ certify SOS).

How can all matrices Q giving $p = z^T Q z$ be parameterized?

Let $w(x)$ be the vector of all monomials of degree $\leq 2d$. For each $Q = Q^T$ there is a unique c such that $z^T Q z = c^T w$; moreover c is a linear function of Q. Hence this association defines a linear mapping $\mathcal{L}$ where $\mathcal{L}(Q) = c$. The domain of $\mathcal{L}$ (the space of symmetric matrices) has dimension $l_{[n,d]}(l_{[n,d]} + 1)/2$, while the range (column vectors) has dimension $l_{[n,2d]}$. Clearly $\mathcal{L}$ has full rank, since any vector c is in the range of $\mathcal{L}$. Therefore, the nullspace of $\mathcal{L}$ has dimension

$$K := \frac{l_{[n,d]}(l_{[n,d]} + 1)}{2} - l_{[n,2d]}$$

and there exist symmetric matrices $\{N_j\}_{j=1}^K$ which form a basis for all N satisfying $z^T N z = 0$. Hence if $p = z^T Q_0 z$, then for all $\lambda_j \in \mathbb{R}$, it also follows that

$$z^T (Q_0 + \sum_{j=1}^K \lambda_j N_j) z = p$$

where the freedom in λ parametrizes all Q with $p = z^T Q z$.

By way of example, for $n = d = 2$, $l_{[n,d]} = 6$, $l_{[n,2d]} = 15$, so $K = 6$. With $z = [1, x_1, x_2, x_1^2, x_1 x_2, x_2^2]^T$, the matrices

$$
N_1 = \begin{bmatrix} 0 & 0 & 0 & 0 & 0 & -1 \\ 0 & 0 & 0 & 0 & 0 & 0 \\ 0 & 0 & 2 & 0 & 0 & 0 \\ 0 & 0 & 0 & 0 & 0 & 0 \\ 0 & 0 & 0 & 0 & 0 & 0 \\ -1 & 0 & 0 & 0 & 0 & 0 \end{bmatrix}, \qquad
N_2 = \begin{bmatrix} 0 & 0 & 0 & 0 & -1 & 0 \\ 0 & 0 & 1 & 0 & 0 & 0 \\ 0 & 1 & 0 & 0 & 0 & 0 \\ 0 & 0 & 0 & 0 & 0 & 0 \\ -1 & 0 & 0 & 0 & 0 & 0 \\ 0 & 0 & 0 & 0 & 0 & 0 \end{bmatrix}
$$

$$
N_3 = \begin{bmatrix} 0 & 0 & 0 & 0 & 0 & 0 \\ 0 & 0 & 0 & 0 & 0 & -1 \\ 0 & 0 & 0 & 0 & 1 & 0 \\ 0 & 0 & 0 & 0 & 0 & 0 \\ 0 & 0 & 1 & 0 & 0 & 0 \\ 0 & -1 & 0 & 0 & 0 & 0 \end{bmatrix}, \qquad
N_4 = \begin{bmatrix} 0 & 0 & 0 & -1 & 0 & 0 \\ 0 & 2 & 0 & 0 & 0 & 0 \\ 0 & 0 & 0 & 0 & 0 & 0 \\ -1 & 0 & 0 & 0 & 0 & 0 \\ 0 & 0 & 0 & 0 & 0 & 0 \\ 0 & 0 & 0 & 0 & 0 & 0 \end{bmatrix}
$$

$$
N_5 = \begin{bmatrix} 0 & 0 & 0 & 0 & 0 & 0 \\ 0 & 0 & 0 & 0 & -1 & 0 \\ 0 & 0 & 0 & 1 & 0 & 0 \\ 0 & 0 & 1 & 0 & 0 & 0 \\ 0 & -1 & 0 & 0 & 0 & 0 \\ 0 & 0 & 0 & 0 & 0 & 0 \end{bmatrix}, \qquad
N_6 = \begin{bmatrix} 0 & 0 & 0 & 0 & 0 & 0 \\ 0 & 0 & 0 & 0 & 0 & 0 \\ 0 & 0 & 0 & 0 & 0 & 0 \\ 0 & 0 & 0 & 0 & 0 & -1 \\ 0 & 0 & 0 & 0 & 2 & 0 \\ 0 & 0 & 0 & -1 & 0 & 0 \end{bmatrix}
$$

form the basis described above. For $q(x_1, x_2)$, a suitable choice for Q_0 is

$$
Q_0 = \begin{bmatrix} 0 & 0 & 0 & 0 & 0 & 0 \\ 0 & 1 & 0 & 0 & 0 & -1 \\ 0 & 0 & 0 & 0 & 0 & 0 \\ 0 & 0 & 0 & 2 & 1 & 0 \\ 0 & 0 & 0 & 1 & -1 & 0 \\ 0 & -1 & 0 & 0 & 0 & 6 \end{bmatrix}.
$$

Note that $Q_0 \nsucceq 0$, but $Q_0 + 6N_6 \succeq 0$. Moreover,

$$
Q_0 + 6N_6 = \begin{bmatrix} 0 & 1 & 0 & 0 & 0 & -1 \\ 0 & 0 & 0 & 2 & 1 & -3 \\ 0 & 0 & 0 & 0 & 3 & 1 \end{bmatrix}^T \begin{bmatrix} 1 & 0 & 0 \\ 0 & \frac{1}{2} & 0 \\ 0 & 0 & \frac{1}{2} \end{bmatrix} \begin{bmatrix} 0 & 1 & 0 & 0 & 0 & -1 \\ 0 & 0 & 0 & 2 & 1 & -3 \\ 0 & 0 & 0 & 0 & 3 & 1 \end{bmatrix},
$$

which illustrates the SOS decomposition given earlier.

Summarizing, given $p \in \mathbb{R}[x]$, there exists a matrix Q_0 (that depends on p) and matrices $\{N_j\}_{j=1}^K$ (these only depend on n and d, and *not* on p) such that

$$
p \text{ is SOS} \quad \Leftrightarrow \quad \exists \lambda \in \mathbb{R}^K \text{ such that } Q_0 + \sum_{j=1}^K \lambda_j N_j \succeq 0
$$

Moreover, if the semidefinite program is infeasible, then the dual variables provide a proof that p is not SOS.

From "checking SOS" to "synthesizing an SOS"

Synthesizing an SOS is necessary when searching for a storage function and/or adjusting parameters in a supply rate to establish dissipativity. Suppose $p_0, p_1, \ldots, p_m \in \mathbb{R}[x]$, with $\partial(p_i) \leq 2d$ for all $i = 0, 1, \ldots, m$. Then regardless of $a \in \mathbb{R}^m$, it follows that $\partial(p_0 + a_1 p_1 + \cdots + a_m p_m) \leq 2d$. The SOS synthesis question is:
When is there a choice of $a \in \mathbb{R}^m$ such that $p_0 + a_1 p_1 + \cdots + a_m p_m$ is SOS in x?

Applying the ideas established thus far we conclude that there exist matrices $\{Q_t\}_{t=0}^m$ (each individually dependent on p_t) and $\{N_j\}_{j=1}^K$ (dependent only on n and d) such that the SOS synthesis is possible if and only if there exist $a \in \mathbb{R}^m$ and $\lambda \in \mathbb{R}^K$ satisfying

$$Q_0 + \sum_{t=1}^m a_t Q_t + \sum_{j=1}^K \lambda_j N_j \succeq 0.$$

An *SOS Program* is an optimization problem that takes this idea one step further, allowing for multiple SOS constraints and a linear objective function. Specifically, a standard form SOS program is given by

$$\begin{aligned}
\underset{a \in \mathbb{R}^m}{\text{minimize}} \quad & c^T a \\
\text{subject to} \quad & f_{1,0}(x) + a_1 f_{1,1}(x) + \cdots + a_m f_{1,m}(x) \in \Sigma[x] \\
& \qquad\qquad\qquad\qquad \vdots \\
& f_{W,0}(x) + a_1 f_{W,1}(x) + \cdots + a_m f_{W,m}(x) \in \Sigma[x]
\end{aligned}$$

where $c \in \mathbb{R}^m$ and $\{f_{b,t}\} \in \mathbb{R}[x]$, $1 \leq b \leq W$, $0 \leq t \leq m$.

Software packages that convert SOS programs into SDPs are available [3–5]. These packages call available SDP solvers, and then convert the results back into polynomial form.

References

1. Choi, M., Lam, T., Reznick, B.: Sums of squares of real polynomials. Proc. Symp. Pure Math. **58**(2), 103–126 (1995)
2. Parillo, P.: Structured semidefinite programs and semialgebraic geometry methods in robustness and optimization. Ph.D. dissertation, California Institute of Technology (2000)
3. Löfberg, J.: Yalmip : a toolbox for modeling and optimization in MATLAB. In: Proceedings of the CACSD Conference, Taipei, Taiwan (2004)
4. Papachristodoulou, A., Anderson, J., Valmorbida, G., Prajna, S., Seiler, P., Parrilo, P.A.: SOS-TOOLS: sum-of squares optimization toolbox for MATLAB. http://arxiv.org/abs/1310.4716 (2013)
5. Seiler, P.: SOSOPT: a toolbox for polynomial optimization. In: arXiv:1308.1889. http://www.aem.umn.edu/~AerospaceControl (2013)

Appendix B
Semidefinite Programming (SDP)

A semidefinite program (SDP) in *inequality form* consists of a linear objective subject to a linear matrix inequality (LMI) constraint:

$$\underset{z\in\mathbb{R}^q}{\text{minimize}} \quad c^T z$$

$$\text{subject to} \quad \sum_{i=1}^{q} z_i A_i - B \geq 0. \tag{B.1}$$

The problem data are the vector $c \in \mathbb{R}^q$ and symmetric matrices $B \in \mathbb{R}^{r\times r}, A_i \in \mathbb{R}^{r\times r}$.

An alternate formulation is the *conic form* which consists of a linear objective, linear constraints, and a matrix decision variable constrained to be positive semidefinite:

$$\underset{X\in\mathbb{R}^{n\times n}}{\text{minimize}} \quad \text{Tr}(GX)$$

$$\text{subject to} \quad \text{Tr}(F_i X) = e_i \quad \text{for } i = 1, \ldots, m \tag{B.2}$$

$$X \geq 0.$$

The problem data are the vector $e \in \mathbb{R}^m$ and symmetric matrices $G \in \mathbb{R}^{n\times n}$, $F_i \in \mathbb{R}^{n\times n}$. The LMI and conic forms are equivalent, in the sense that one can be converted into the other by introducing new variables and constraints. For notational simplicity we will refer to the conic form SDP in the remainder of this section.

Standard SDP solvers [1–3] use primal-dual interior point algorithms. These algorithms have worst-case polynomial complexity [4] but can become computationally intractable for large problems. The computational complexity depends on the number of constraints m, the dimension of the semidefinite cone n, and the structure and sparsity of the problem data.

While most solvers automatically take advantage of the sparsity in the problem data, additional approaches have been developed to exploit further structure in the

© The Author(s) 2016
M. Arcak et al., *Networks of Dissipative Systems*,
SpringerBriefs in Control, Automation and Robotics,
DOI 10.1007/978-3-319-29928-0

problem. For SDPs with symmetry in the problem data it was shown in [5] that both the dimension and the number of constraints can be reduced. References [6, 7] consider SDPs that have a chordal sparsity pattern in the problem data. This allows the LMI constraint to be reduced to multiple smaller LMIs without adding conservatism.

Another approach to improving the scalability of SDPs, proposed in [8, 9], is to constrain the decision matrix X to an inner approximation of the cone of positive semidefinite matrices. Although this introduces conservatism, depending on the approximation, it can improve the computational efficiency significantly. References [8, 9] propose two approximations that achieve this goal: the diagonally-dominant (DD) and scaled diagonally-dominant (SDD) cones of symmetric matrices.

Definition B.1 The cone of real symmetric DD matrices with nonnegative diagonal entries is

$$\mathbb{S}^n_{DD} = \left\{ X = X^T \in \mathbb{R}^{n \times n} : x_{ii} \geq \sum_{j \neq i} |x_{ij}| \text{ for all } i \right\}.$$

Real symmetric DD matrices with nonnegative diagonal entries are positive semidefinite by Gershgorin's disc criterion:

Theorem B.1 *Let $X \in \mathbb{R}^{n \times n}$ and $D(x_{ii}, R_i)$ be the closed disk centered at x_{ii} with radius $R_i = \sum_{j \neq i} |x_{ij}|$. Every eigenvalue of X is contained in at least one disk $D(x_{ii}, R_i)$.*

The set of DD matrices is characterized by linear constraints. Therefore, replacing the constraint $X \geq 0$ in (B.2) with $X \in \mathbb{S}^n_{DD}$ gives a linear program (LP).

Definition B.2 The cone of symmetric SDD matrices is

$$\mathbb{S}^n_{SDD} = \left\{ X = X^T \in \mathbb{R}^{n \times n} : \exists \text{ a positive diagonal } S \in \mathbb{R}^{n \times n} \text{ s.t. } SXS \in \mathbb{S}^n_{DD} \right\}.$$

Clearly, $\mathbb{S}^n_{DD}$ is a subset of $\mathbb{S}^n_{SDD}$. For a positive diagonal matrix $S \in \mathbb{R}^{n \times n}$ and $X \in \mathbb{R}^{n \times n}$ the eigenvalues of X and SXS are the same, so SDD matrices are also positive semidefinite.

Let $M^{ij} \in \mathbb{R}^{n \times n}$ denote the symmetric matrix where the only nonzero entries are $m_{ii}, m_{ij}, m_{ji},$ and m_{jj}. In [8] it is shown that the cone of symmetric SDD matrices of dimension n can be characterized as

$$\mathbb{S}^n_{SDD} = \left\{ X = X^T \in \mathbb{R}^{n \times n} : X = \sum_{i=1}^{n} \sum_{j>i}^{n} M^{ij}, \begin{bmatrix} m_{ii} & m_{ij} \\ m_{ji} & m_{jj} \end{bmatrix} \geq 0 \text{ for all } i, j > i \right\}.$$

Since the matrices constrained to be positive semidefinite are of dimension two, $M^{ij} \geq 0$ is equivalent to

$$m_{ii} \geq 0, \quad m_{jj} \geq 0, \quad m_{ii} m_{jj} \geq m_{ij}^2.$$

Therefore, replacing $X \geq 0$ in (B.2) with $X \in \mathbb{S}^n_{SDD}$ gives a second order cone program (SOCP) [10].

The DD or SDD cone of matrices are strict subsets of the cone of semidefinite matrices. Therefore, restricting the LMI to be DD or SDD introduces conservatism, but solvers for LP and SOCP problems are much more efficient and scalable than standard SDP solvers.

SDP Duality

Primal-dual algorithms, used by most SDP solvers, simultaneously attempt to solve the primal problem, (B.1) or (B.2), and the corresponding dual problem. The dual problem of the inequality form SDP is

$$\underset{Z \in \mathbb{R}^{r \times r}}{\text{maximize}} \quad \text{Tr}(BZ)$$
$$\text{subject to} \quad \text{Tr}(A_i Z) = c_i \quad \text{for } i = 1, \ldots, q \tag{B.3}$$
$$Z \geq 0$$

where A_i, B, and c are the same as in (B.1) and $Z \in \mathbb{R}^{r \times r}$ is the dual variable. For the conic form SDP the dual problem is

$$\underset{x \in \mathbb{R}^q}{\text{maximize}} \quad e^T x$$
$$\text{subject to} \quad \sum_{i=1}^{q} x_i F_i - G \leq 0 \tag{B.4}$$

where F_i, G, and e are the same as in (B.2) and $x \in \mathbb{R}^q$ is the dual variable.

We denote the optimal value of the primal problem as $p = c^T z^\star = \text{Tr}(GX^\star)$ where $z^\star$ and $X^\star$ are the optimal solutions of (B.1) and (B.2), respectively. Similarly, we denote the optimal value of the dual problem as $d = \text{Tr}(BZ^\star) = e^T x^\star$ where $x^\star$ and $Z^\star$ are the optimal solutions of (B.3) and (B.4), respectively.

Weak duality ($d \leq p$) holds for any SDP. If $d = p$ it is said that strong duality holds. For LPs strong duality always holds, but this is not the case for general SDPs. By Slater's condition, strong duality holds if the primal and dual problems are strictly feasible. If strong duality does not hold primal-dual SDP solvers may return inaccurate solutions. Therefore, it is a good idea to check that the returned solution is reasonable and satisfies the problem constraints.

When No Strictly Feasible Solution Exists

When a strictly feasible solution does not exist, SDP solvers require more computational time and may yield inaccurate solutions. The reasons for this are that the problem is larger than necessary (i.e., it can be reformulated as an equivalent, but lower dimension SDP) and strong duality may not hold. For example, certifying the passivity of a linear system requires finding $P \geq 0$ such that

$$\begin{bmatrix} A^T P + PA & PB - C^T \\ B^T P - C & 0 \end{bmatrix} \leq 0 \tag{B.5}$$

which is not strictly feasible, i.e., it cannot hold with strict inequality. In addition, (B.5) implicitly contains the equality constraint $PB = C^T$.

In cases where there are implicit equality constraints, it may be possible to reformulate the problem in an equivalent form. A reformulation for (B.5) is

$$A^T P + PA \leq 0 \tag{B.6}$$

$$PB = C^T. \tag{B.7}$$

Although this is mathematically equivalent, it is much easier for SDP solvers to attain an accurate solution when the equality constraint is explicitly specified and the LMI constraint is strictly feasible.

However, in general it is not obvious how to manually reformulate the problem. In [11] an efficient computational method was developed to automatically detect problems with no strictly feasible solution and to reformulate the problem with a preprocessing procedure.

References

1. ApS, M.: The MOSEK optimization toolbox for MATLAB manual. Version 7.1 (Revision 28). http://docs.mosek.com/7.1/toolbox/index.html (2015)
2. Sturm, J.F.: Using SeDuMi 1.02, a Matlab toolbox for optimization over symmetric cones. Optim. Methods Softw. 11(1–4), 625–653 (1999)
3. Toh, K., Todd, M., Tutuncu, R.: SDPT3—a Matlab software package for semidefinite programming. Optim. Methods Softw. 11, 545–581 (1999)
4. Vanderberghe, L., Boyd, S.: Semidefinite programming. SIAM Rev. 38(1), 49–95 (1996)
5. Gatermann, K., Parrilo, P.A.: Symmetry groups, semidefinite programs, and sums of squares. J. Pure Appl. Algebra 192(1–3), 95–128 (2004)

6. Khoshfetrat Pakazad, S., Hansson, A., Andersen, M.S., Rantzer, A.: Distributed robustness analysis of interconnected uncertain systems using chordal decomposition. ArXiv e-prints (2014)
7. Mason, R., Papachristodoulou, A.: Chordal sparsity, decomposing SDPs and the Lyapunov equation. In: American Control Conference (ACC), pp. 531–537 (2014). doi:10.1109/ACC.2014.6859255
8. Ahmadi, A.A., Majumdar, A.: DSOS and SDSOS optimization: LP and SOCP-based alternatives to sum-of squares optimization. In: Conference on Information Sciences and Systems (2014)
9. Majumdar, A., Ahmadi, A., Tedrake, R.: Control and verification of high-dimensional systems with DSOS and SDSOS programming. In: 2014 IEEE 53rd Annual Conference on Decision and Control (CDC), pp. 394–401 (2014)
10. Alizadeh, F., Goldfarb, D.: Second-order cone programming. Math. Program. **95**(1), 3–51 (2003). doi: 10.1007/s10107-002-0339-5. http://dx.doi.org/10.1007/s10107-002-0339-5
11. Permenter, F., Parrilo, P.: Partial facial reduction: simplified, equivalent SDPs via approximations of the PSD cone. ArXiv e-prints (2014)

Appendix C
The KYP Lemma

The following result, quoted from [1], is a streamlined version of the classical KYP Lemma due to Kalman [2], Yakubovich [3], and Popov [4].

Theorem C.1 *Given $F \in \mathbb{R}^{n \times n}$, $G \in \mathbb{R}^{n \times m}$, $\Gamma = \Gamma^T \in \mathbb{R}^{(n+m) \times (n+m)}$ with $\det(j\omega I - F) \neq 0 \ \forall \omega \in \mathbb{R}$ and (F, G) controllable, the following statements are equivalent:*
(1) For all $\omega \in \mathbb{R} \cup \{\infty\}$,

$$\begin{bmatrix} (j\omega I - F)^{-1} G \\ I \end{bmatrix}^* \Gamma \begin{bmatrix} (j\omega I - F)^{-1} G \\ I \end{bmatrix} \leq 0. \tag{C.1}$$

(2) There exists $P = P^T \in \mathbb{R}^{n \times n}$ such that

$$\begin{bmatrix} F^T P + PF & PG \\ G^T P & 0 \end{bmatrix} + \Gamma \leq 0. \tag{C.2}$$

The corresponding equivalence for strict inequalities holds even if (F, G) is not controllable. In addition, if F is Hurwitz (all eigenvalues have negative real parts) and the upper left corner of Γ is positive semidefinite, then $P \geq 0$.

Example C.1 Consider the system (8.24) in Example 8.3. To show that a block separable Lyapunov function

$$V_1(x_1, x_2) + V_2(x_3) = \begin{bmatrix} x_1 & x_2 \end{bmatrix}^T P_1 \begin{bmatrix} x_1 \\ x_2 \end{bmatrix} + p_2 x_3^2$$

© The Author(s) 2016
M. Arcak et al., *Networks of Dissipative Systems*,
SpringerBriefs in Control, Automation and Robotics,
DOI 10.1007/978-3-319-29928-0

does not exist we suppose, to the contrary, there exist $P_1 = P_1^T \in \mathbb{R}^{2\times2}$ and scalar $p_2 > 0$ such that

$$\begin{bmatrix} P_1 & 0 \\ 0 & p_2 \end{bmatrix} \begin{bmatrix} 0 & 1 & 0 \\ -1 & -0.5 & -6 \\ 1 & 1 & -0.6 \end{bmatrix} + \begin{bmatrix} 0 & 1 & 0 \\ -1 & -0.5 & -6 \\ 1 & 1 & -0.6 \end{bmatrix}^T \begin{bmatrix} P_1 & 0 \\ 0 & p_2 \end{bmatrix} \le 0. \tag{C.3}$$

Since $p_2 > 0$ can be factored out we set $p_2 = 1$ without loss of generality. We define

$$F = \begin{bmatrix} 0 & 1 \\ -1 & -0.5 \end{bmatrix} \quad G = \begin{bmatrix} 0 \\ -6 \end{bmatrix} \quad H = \begin{bmatrix} 1 & 1 \end{bmatrix},$$

drop the subscript from P_1, and rewrite (C.3) as

$$\begin{bmatrix} F^T P + PF & PG \\ G^T P & 0 \end{bmatrix} + \Gamma \le 0 \quad \text{where} \quad \Gamma = \begin{bmatrix} 0 & H^T \\ H & -1.2 \end{bmatrix}. \tag{C.4}$$

Since (F, G) is controllable and $\det(j\omega I - F) = (1 - \omega^2) + j(0.5\omega) \ne 0 \ \forall \omega \in \mathbb{R}$, Theorem C.1 states that (C.4) is equivalent to

$$H(j\omega I - F)^{-1}G + (H(j\omega I - F)^{-1}G)^* - 1.2 \le 0 \quad \forall \omega \in \mathbb{R} \cup \{\infty\}, \tag{C.5}$$

which means $Re\{H(j\omega I - F)^{-1}G\} \le 0.6$. However, for $\omega^2 \in (2.75, 4)$,

$$Re\{H(j\omega I - F)^{-1}G\} = Re\left\{-6\frac{1 + j\omega}{(1 - \omega^2) + j(0.5\omega)}\right\} = -6\frac{1 - 0.5\omega^2}{\omega^4 - 1.75\omega^2 + 1} > 0.6$$

thus contradicting the hypothesis that there exist $P_1 = P_1^T \in \mathbb{R}^{2\times2}$ and $p_2 > 0$ satisfying (C.3).

References

1. Rantzer, A.: On the Kalman-Yakubovich-Popov lemma. Syst. Control Lett. **28**, 7–10 (1996)
2. Kalman, R.: Canonical structure of linear dynamical systems. Proc. Natl. Acad. Sci. U.S.A. **48**, 596–600 (1962)
3. Yakubovich, V.: The solution of certain matrix inequalities in automatic control theory. Dokl. Akad. Nauk **143**, 1304–1307 (1962)
4. Popov, V.: The solution of a new stability problem for controlled systems. Autom. Remote Control **24**, 1–23 (1963); Translated from Avtomatika i Telemekhanika **24**, 7–26 (1963)

Appendix D
True/False Questions for Chapter 1

1. Suppose the function $h : \mathbb{R}^n \times \mathbb{R}^m \to \mathbb{R}^p$ in (1.2) is invertible (with $p = m$) in the sense that for all $x \in \mathbb{R}^n$, $y \in \mathbb{R}^p$, there is a unique $u \in \mathbb{R}^m$ such that $h(x, u) = y$. Denote this u as $h_I(x, y)$, where $h_I : \mathbb{R}^n \times \mathbb{R}^p \to \mathbb{R}^m$. Define the inverse system (with input v, output w, and state η)

$$\frac{d}{dt}\eta(t) = f(\eta(t), h_I(\eta(t), v(t))), \quad w(t) = h_I(\eta(t), v(t)) \qquad (D.1)$$

and note that for any $\xi \in \mathbb{R}^n$, (u, y) solves (1.1)–(1.2) with $x(0) = \xi$ if and only if $v = y$, $w = u$ solves (D.1) with $\eta(0) = \xi$.
True/False: The system in (1.1)–(1.2) is dissipative with respect to the supply rate $s(u, y)$ if and only if the inverse system is dissipative with respect to $\hat{s}(v, w) := s(w, v)$.

2. **True/False**: If a dynamical system G is dissipative with respect to supply rates s_1 and s_2, then it is dissipative with respect to the supply rate $s(u, y) := s_1(u, y) - s_2(u, y)$.

3. **True/False**: If a dynamical system G is dissipative with respect to supply rates s_1 and s_2, then it is dissipative with respect to the supply rate $s(u, y) := \alpha s_1(u, y) + (1 - \alpha)s_2(u, y)$ for all $0 \le \alpha \le 1$.

4. **True/False**: If a dynamical system G is dissipative with respect to supply rates s_1 and s_2, then it is dissipative with respect to the supply rate $s(u, y) := \alpha s_1(u, y) + \beta s_2(u, y)$ for all $\alpha \ge 0$, $\beta \ge 0$.

5. For a dynamical system G, let $-G$ denote the same system with the sign of the output reversed.
True/False: G is dissipative with respect to s if and only if $-G$ is dissipative with respect to $-s$.

6. The "sum" of two dynamical systems G_1 and G_2 is a dynamical system defined by $y = G_1(u) + G_2(u)$.
True/False: If G_i is dissipative with respect to $s_i(u_i, y_i)$, $i = 1, 2$, then the sum $G_1 + G_2$ is dissipative with respect to $s(u, y) := s_1(u, y) + s_2(u, y)$.

© The Author(s) 2016
M. Arcak et al., *Networks of Dissipative Systems*,
SpringerBriefs in Control, Automation and Robotics,
DOI 10.1007/978-3-319-29928-0

7. **True/False**: If each G_i is dissipative with respect to $u_i^T y_i$, then the sum $G_1 + G_2$ is dissipative with respect to $s(u, y) := u^T y$.

8. Given a scalar $d > 0$, define a dynamical system G_d as $G_d := d \circ G \circ d^{-1}$. Let u and y denote the input and output of G, and v and w denote the input and output of G_d, so that $w = dG(v/d)$. Note that if G is nonlinear, then in general, $G_d \neq G$. **True/False**: G is dissipative with respect to a *quadratic* supply rate $s(u, y)$ if and only if G_d is dissipative with respect to $s(v, w)$.

9. **True/False**: Let

$$\frac{\mathrm{d}}{\mathrm{d}t} x(t) = f(x(t), u(t)), \quad y(t) = h(x(t), u(t))$$

describe a nonlinear dynamical system G. For every $\alpha > 0$, the dynamical system (with input v, output w, and state η)

$$\frac{\mathrm{d}}{\mathrm{d}t} \eta(t) = \alpha f(\eta(t), v(t)), \quad w(t) = h(\eta(t), v(t))$$

is dissipative with respect to exactly the same supply rates as G.

10. **True/False**: Suppose $W \in \mathbb{R}^{(m+p)\times(m+p)}$ has $W = W^T \succeq 0$. Every dynamical system (with appropriate input and output dimension) is dissipative with respect to the supply rate

$$s(u, y) := \begin{bmatrix} u \\ y \end{bmatrix}^T W \begin{bmatrix} u \\ y \end{bmatrix}.$$

11. **True/False**: If the dynamical system G is dissipative with respect to the quadratic supply rate s, then for every $\alpha \in [0, 1]$, the dynamical system αG (output scaled by α) is dissipative with respect to s.

Answers: 1:T, 2:F, 3:T, 4:T, 5:F, 6:F, 7:T, 8:T, 9:T, 10:T, 11:F

Series Editors' Biographies

Tamer Başar is with the University of Illinois at Urbana-Champaign, where he holds the academic positions of Swanlund Endowed Chair, Center for Advanced Study Professor of Electrical and Computer Engineering, Research Professor at the Coordinated Science Laboratory, and Research Professor at the Information Trust Institute. He received the B.S.E.E. degree from Robert College, Istanbul, and the M.S., M.Phil, and Ph.D. degrees from Yale University. He has published extensively in systems, control, communications, and dynamic games, and has current research interests that address fundamental issues in these areas along with applications such as formation in adversarial environments, network security, resilience in cyber-physical systems, and pricing in networks.

In addition to his editorial involvement with these Briefs, Basar is also the Editor-in-Chief of Automatica, Editor of two Birkhäuser Series on Systems and Control and Static and Dynamic Game Theory, the Managing Editor of the Annals of the International Society of Dynamic Games (ISDG), and member of editorial and advisory boards of several international journals in control, wireless networks, and applied mathematics. He has received several awards and recognitions over the years, among which are the Medal of Science of Turkey (1993); Bode Lecture Prize (2004) of IEEE CSS; Quazza Medal (2005) of IFAC; Bellman Control Heritage Award (2006) of AACC; and Isaacs Award (2010) of ISDG. He is a member of the US National Academy of Engineering, Fellow of IEEE and IFAC, Council Member of IFAC (2011–2014), a past president of CSS, the founding president of ISDG, and president of AACC (2010–2011).

Antonio Bicchi is Professor of Automatic Control and Robotics at the University of Pisa. He graduated from the University of Bologna in 1988 and was a postdoc scholar at M.I.T. A.I. Lab between 1988 and 1990. His main research interests are in:

© The Author(s) 2016
M. Arcak et al., *Networks of Dissipative Systems*,
SpringerBriefs in Control, Automation and Robotics,
DOI 10.1007/978-3-319-29928-0

- dynamics, kinematics and control of complex mechanical systems, including robots, autonomous vehicles, and automotive systems;
- haptics and dextrous manipulation; and theory and control of nonlinear systems, in particular hybrid (logic/dynamic, symbol/signal) systems.
- theory and control of nonlinear systems, in particular hybrid (logic/dynamic, symbol/signal) systems.

He has published more than 300 papers in international journals, books, and refereed conferences.

Professor Bicchi currently serves as the Director of the Interdepartmental Research Center "E. Piaggio" of the University of Pisa, and President of the Italian Association or Researchers in Automatic Control. He has served as Editor in Chief of the Conference Editorial Board for the IEEE Robotics and Automation Society (RAS), and as Vice President of IEEE RAS, Distinguished Lecturer, and Editor for several scientific journals including the *International Journal of Robotics Research, the IEEE Transactions on Robotics and Automation, and IEEE RAS Magazine*. He has organized and co-chaired the first WorldHaptics Conference (2005), and Hybrid Systems: Computation and Control (2007). He is the recipient of several best paper awards at various conferences, and of an Advanced Grant from the European Research Council. Antonio Bicchi has been an IEEE Fellow since 2005.

Miroslav Krstic holds the Daniel L. Alspach chair and is the founding director of the Cymer Center for Control Systems and Dynamics at University of California, San Diego. He is a recipient of the PECASE, NSF Career, and ONR Young Investigator Awards, as well as the Axelby and Schuck Paper Prizes. Professor Krstic was the first recipient of the UCSD Research Award in the area of engineering and has held the Russell Severance Springer Distinguished Visiting Professorship at UC Berkeley and the Harold W. Sorenson Distinguished Professorship at UCSD. He is a Fellow of IEEE and IFAC. Professor Krstic serves as Senior Editor for *Automatica and IEEE Transactions on Automatic Control* and as Editor for the Springer series *Communications and Control Engineering*. He has served as Vice President for Technical Activities of the IEEE Control Systems Society. Krstic has co-authored eight books on adaptive, nonlinear, and stochastic control, extremum seeking, control of PDE systems including turbulent flows and control of delay systems.

Index

© The Author(s) 2016
M. Arcak et al., *Networks of Dissipative Systems*,
SpringerBriefs in Control, Automation and Robotics,
DOI 10.1007/978-3-319-29928-0